C0-FYN-641

PRENTICE-HALL

Foundations of Economic Geography Series

NORTON GINSBURG, *Editor*

GEOGRAPHY OF MANUFACTURING, *Gunnar Alexandersson*

GEOGRAPHY OF MARKET CENTERS AND RETAIL DISTRIBUTION, *Brian J. L. Berry*

GEOGRAPHY OF AGRICULTURE, *Howard F. Gregor*

ENERGY IN THE PERSPECTIVE OF GEOGRAPHY, *Nathaniel B. Guyol*

MANUFACTURING LOCATION, *James Lindberg*

GEOGRAPHY OF URBAN LAND USE, *Harold Mayer*

GEOGRAPHY OF WATER RESOURCES, *W. R. Derrick Sewell*

GEOGRAPHY OF TRANSPORTATION, *Edward J. Taaffe* and *Howard L. Gauthier, Jr.*

GEOGRAPHY OF INTERNATIONAL TRADE, *Richard S. Thoman* and *Edgar C. Conkling*

THE MERCHANT'S WORLD, *James E. Vance, Jr.*

A PROLOGUE TO POPULATION GEOGRAPHY, *Wilbur Zelinsky*

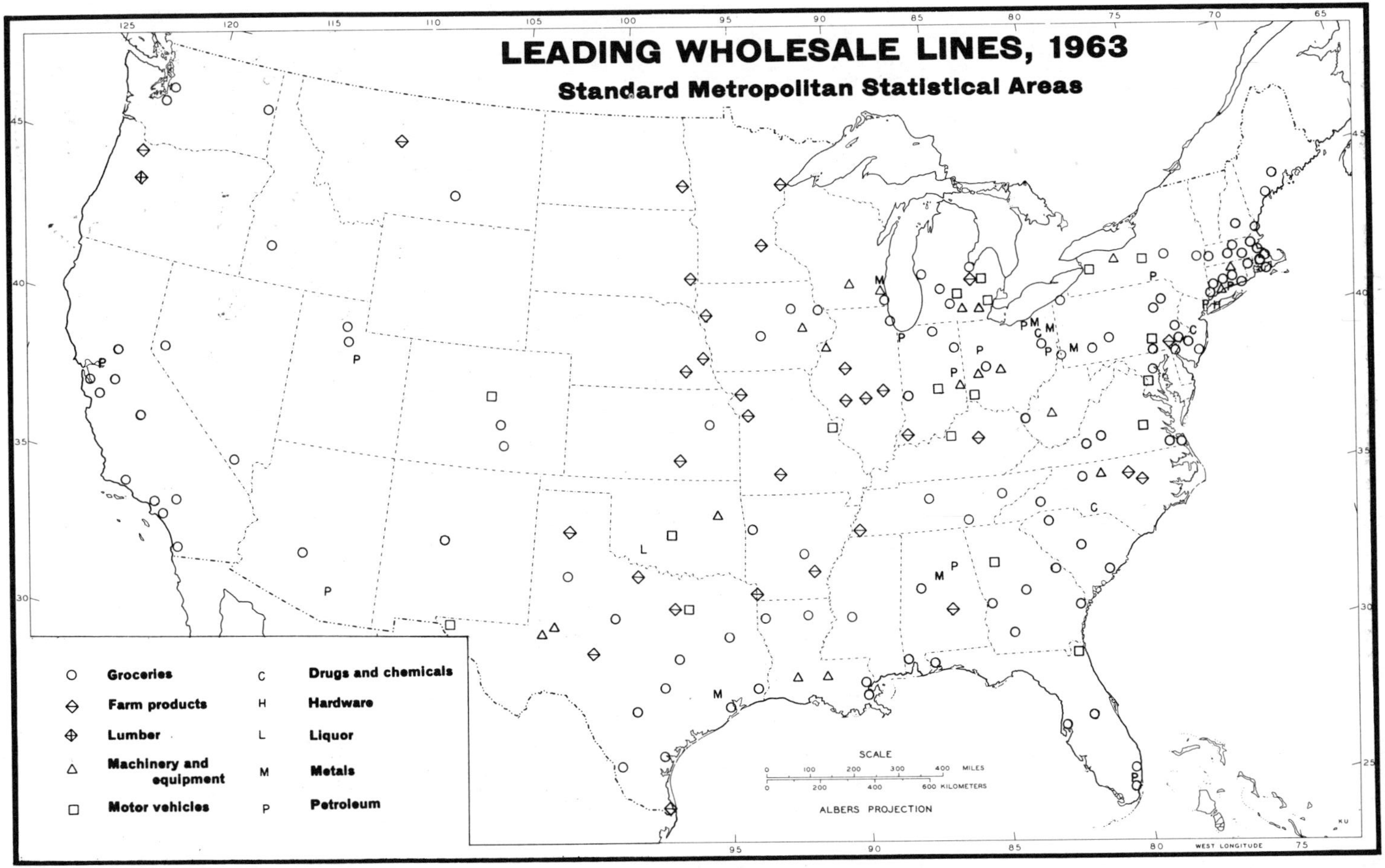

LEADING WHOLESALE LINES, 1963
Standard Metropolitan Statistical Areas
Groceries
Farm products
Lumber
Machinery and equipment
Motor vehicles
C Drugs and chemicals
H Hardware
L Liquor
M Metals
P Petroleum
SCALE
0 100 200 300 400 MILES
0 200 400 600 KILOMETERS
ALBERS PROJECTION
WEST LONGITUDE
KU

The Merchant's World

THE GEOGRAPHY OF WHOLESALING

JAMES E. VANCE, JR.

Professor of Geography
University of California, Berkeley

Theodore Lownik Library
Illinois Benedictine College
LISLE, IL 60532

[redacted]

PRENTICE-HALL, INC., Englewood Cliffs, N.J.

658.86
V277m

© *Copyright 1970 by Prentice-Hall, Inc.*
Englewood Cliffs, N.J.

All rights reserved. No part of this book may be reproduced in any form or by any means without permission in writing from the publisher. Printed in the United States of America.

Library of Congress Catalog Card No.: 73–132041

13–577114–5

Current printing (last number):

10 9 8 7 6 5 4 3 2 1

PRENTICE-HALL INTERNATIONAL, INC., *London*
PRENTICE-HALL OF AUSTRALIA, PTY. LTD., *Sydney*
PRENTICE-HALL OF CANADA, LTD., *Toronto*
PRENTICE-HALL OF INDIA PRIVATE LTD., *New Delhi*
PRENTICE-HALL OF JAPAN, INC., *Tokyo*

Foundations of Economic Geography Series

Among the various fields of geography, economic geography, perhaps more than any other, has experienced remarkable changes within the past twenty years—so many that it is almost impossible for one scholar to command all aspects of it. The result has been increasing specialization on the one hand and, on the other, a fundamental need for bringing the fruits of that specialization to students of economic geography.

The *Foundations of Economic Geography* Series consists of several volumes, each focusing on a major problem in economic geography. It is designed to bring the student, whether novice or more experienced, to the frontiers of knowledge in economic geography, and in so doing, forcefully to demonstrate the methodological implications of current research—but at a level comprehensible even to those just becoming aware of the fascinating problems in the field as it is developing today.

Each volume stands as a contribution to understanding in its own right, but the series as a whole is intended to provide a broad cross-section of on-going research in economic geography, stemming from concern with a variety of problems. On the other hand, the series should not be regarded as a complete synthesis of work in economic geography, although the volumes explore in depth certain major issues of keenest interest to economic geographers and others in related fields to a degree impossible in textbooks that attempt to cover the entire field. At the same time, the student is brought face-to-face with the kinds of intellectual and conceptual problems that characterize economic geography in a way that no over-all survey can accomplish. Each volume thus provides a basis for an intensive exploration of issues that constitute the cutting edge of research in this most dynamic and demanding field of knowledge.

As time goes on and new volumes appear in the series, the original volumes will be modified in keeping with new developments and orientations, not only in economic geography, but in the field of geography as a whole. The first volume to appear in the series, Wilbur Zelinsky's *A Prologue to Population Geography*, acts as a bridge between economic and cultural geography and as a means for exploring ideas and methods concerning a problem of increasing interest to geographers and social scientists alike: the growth, diffusion, and distribution of populations throughout the world. Brian J. L. Berry's *Geography of Market Centers and Retail Distribution* attempts to fill a major lacuna in the literature of economic geography, as it lays down principles concerning the spatial distribution and organization of marketing in both advanced and lesser developed economies. In so doing, it provides a bridge between the geographies of consumption, production, and cities, and links them for perhaps the first time effectively through a theoretical system, still primitive, but far in advance of comparable formulations.

Gunnar Alexandersson's *Geography of Manufacturing* reflects the need for considering the historical-ecological settings within which manufacturing enterprises originate and flourish. Though superficially nontheoretical, it contains flashing insights into the extreme socioeconomic complexities that have resulted in the world pattern of manufacturing, and it is concerned with an interpretation of that pattern through an evolutionary-descriptive technique applied to selected industries and regions.

Richard E. Thoman and Edgar C. Conkling in their *Geography of International Trade* deal with a topic that has not been given the attention it deserves by economic geographers. By careful analysis of trade data and an imaginative use of graphic and tabular devices, they interpret the pattern and structure of international trade in terms of current monetary and economic blocs. The result is the first modern treatment of one of the basic types of international relations, and thereby an important contribution to the political as well as economic geographical literature.

In his *Geography of Agriculture: Themes in Research,* Howard Gregor has methodically and imaginatively analyzed the intellectual history of this subdiscipline of economic geography. His bibliography alone is a substantial achievement. Even more important, he has presented with extraordinary clarity the major geographical problems relating to the study of agriculture, all in simple, straightforward prose, but at a level of intellectual discourse of as great interest to professional geographers as to novices in the field. He also relates the results of research in agricultural geography to real problems in the nonacademic world.

Unusual for one of a series of this kind, James E. Vance, Jr.'s *The Merchant's World: The Geography of Wholesaling* makes a substantial original contribution to knowledge through the examination of principles applying to the distribution and organization of wholesaling activities in the past several hundred years. Although readily comprehensible to literate neophytes in economic geography, Professor Vance's thesis suggests a major dimension in the geographic study of economic activity, which has been seriously neglected so far. Moreover, he raises fundamental and intellectually demanding questions about the state of theory in economic geography, and his work illustrates the value of an historical–analytical approach to geographic problems.

The other volumes in the series, whether concerned with transportation, urban land uses, manufactural location, or resources management, display "bridging" qualities that transcend the narrow limitations of ordinary descriptive handbooks. All are concerned with the new and the fresh as well as the traditional, and with the transformation of a somewhat parochial field of scholarship into one that is interdisciplinary as well as innovative and pioneering

Norton Ginsburg

The University of Chicago

Contents

to E. M. V., whose clear perception saw the road

Despite its ultimate reaching back into the middle ages for an understanding of the rôle of wholesaling, this study when first undertaken centered upon the location of wholesale trade in modern American cities. In that first shape it was aided appreciably by a grant from the then Real Estate Research Program of the University of California, and by the enthusiastic searching out of material, and of gaps in my logic, by two assistants, Mr. James Sontag and Mr. Roger Barnett. When the shape began to change to a systemic view of wholesale trade over the course of European settlement in the newer lands it became necessary to turn more to library materials and I was lucky indeed to be able to ask yet again the invaluable assistance of Miss Marion Henderson and the Clark University Library. While doing so my life was made much more pleasant because DeWolf and Kitty Perry still practice New England hospitality, and because Tiffany and Jean cheerfully gambled their future.

JAMES E. VANCE, JR.

CHAPTER 1 *introduction: wholesaling, the pioneer in space*

No one can tell us what man's initial thought was concerning the world in which he found himself, but we may well assume that early in his awareness of the world he came to appreciate that his view of it was incomplete. Space could stretch from the width of a line he drew in the dirt with his finger, through the ultimate range of his vision to the farthest reach of his knowledge, and, for a long time, on into a dark and frightening unknown farther away. Because men have always apprehended space through both their perceptions and their imaginations, sometimes failing to distinguish between the two, it would be misleading to consider the concept of space as merely a sense of distances. The question of whether truths had veracity only within a village or throughout the universe has led man to explore, through both observation and abstract thought, first the world beyond his horizon and, lately, the space beyond the world. Much of the conceptual work in the field of geography has been concerned with discovering how man conceives and then makes use of space: this study continues this search.

Long before man learned the full extent of his own world he was forced to deal with space, in a more limited context, in the effort to shape his individual economic life. In that context several truisms must be noted. Biologically we must assume that existence of the family economic unit from the very *fons et origo* of man's use of the earth; but it is also reasonable to assume that many families, even in the beginning, were not alike in their self-sufficiency, and that quite early in human history this led to attempts to supplement familial provision. *Trade* would have to be introduced, probably first between families in close kinship relation but ultimately among families with no biological ties. Still, a geographical association would logically persist, as early trade required that a producer and a consumer meet to accomplish their exchange. That meeting was the beginning of what we would call retail trade.

We could draw many conclusions from the fact that early trade was conducted face-to-face, but only two need concern us here. The first relates to the locus of early retail trade. In the initial association of producers and consumers the place of sale fell naturally into one of two locales: city workmen were able to meet consumers in their shops, for the customers assembled in town; but rural workmen (agriculturalist) had to devise a special functional entity, divorced from their workshop in the fields, in order to reach their customers. This was the market square, which was universal until recently and still holds great importance in most countries outside of Northwestern Europe and Anglo-America. The second derivative conclusion relates to the nature of the men conducting the retail trade. For the most part we may assume that the craftsmen who sold goods were a part of the same ethnic group that was doing the buying. But in a number of cases—particularly with respect to some specialized trades and in instances of despised occupations—alien peoples might be involved. In that event their place of work might lie beside the city, as a ghetto, urban and retail-oriented, although clearly distinguished in morphological position from the settlement of the natives. But in all these cases the nature of the trade was such that producer and consumer *met in space* to conduct their own transactions. Man's need for spatial knowledge could remain limited.

Up to this point we have assumed a geographical self-sufficiency even in the absence of the material adequacy of family resources. Removing that stricture of local support led quickly to fundamental changes in the personnel and practices of trade, modifications of great importance to our present concern with wholesale trade. In the brief discussion of the basic conditions of retail exchange, among individuals in a localized situation above, we saw that the place of trade and the persons involved were *functionally determinate*. The artisan worked and sold within his own quarter, the farmer or his wife had his stall in the market, and the customers, whose range of movement was restricted, could only buy from local producers. In this situation a single linkage existed and for that reason geographical analysis is simple and predictable. The situation was also *scalar determinate* in the sense that the extent of trade was predetermined by the size of the local population accessible for direct face-to-face exchange. Undoubtedly any transformation in the pattern of trade stemmed from the growth or decline in size and prosperity of the parochial population. This is simply a case of *endogenic change*, which becomes the main dynamic of retail trade.

The notion of endogenic change is the substantial grounding for the main existing theory seeking to account for the interaction of economic activity and settlement pattern, Walter Christaller's formulation of central-place theory.[1] When first proposed in 1933 central-place theory was

[1] Walter Christaller, *Central Places in Southern Germany*, trans. Carlisle Baskin (Englewood Cliffs, N.J.: Prentice-Hall, Inc., 1966).

THE DEVELOPMENT OF COMPLEX SYSTEMS

The Unit

The individual self-sufficient farm.

The shared unit on which either system is predicated.

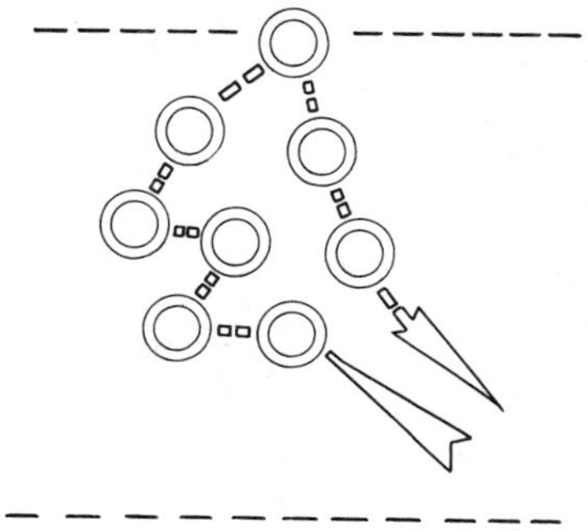

The Circuit

The individual farms loosely tied together by the trading circuit of itinerant and periodic traders (in commodities absent in the locality).

The origin of the open system through its introduction of the practice of agency in trade.

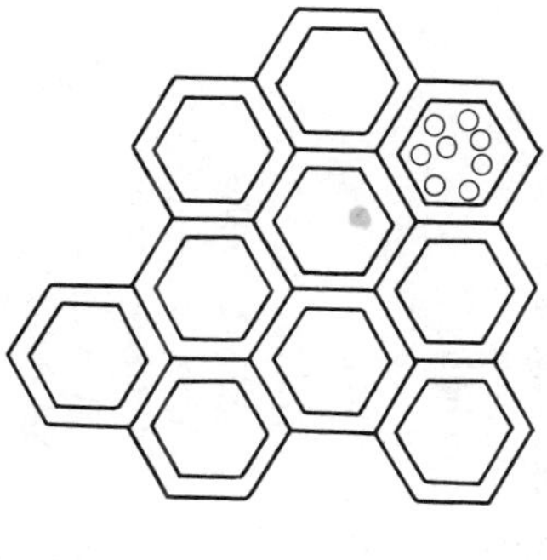

The Closed System

The individual central-place tributary area, with its core settlement.

The origin of the closed system through its introduction of the restriction to parochial economic support through the dependence upon trade-service conflux.

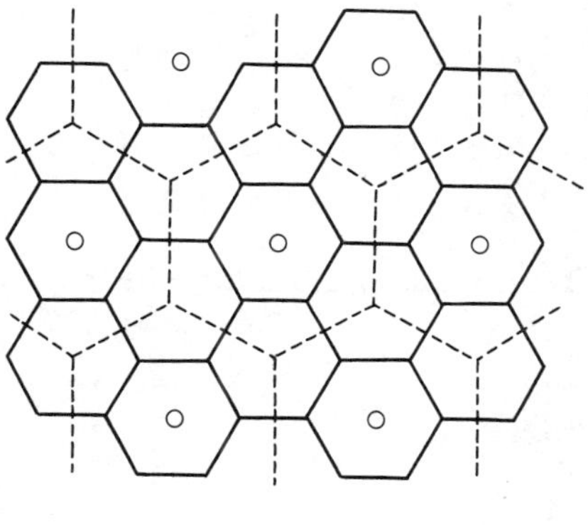

The Central-place System

The central-place system with its hierarchical assembly of internally determined and transformed conflluctive units.

The basis, up to now, of the theory of spatial sorting and ordering of cities.

With the fundamental requirement of geographical uniformity, economic isolation, and historical stratification.

Higher-order places can emerge only with increasing rural productivity and enlarging hierarchical assembly.

Figure 1

THE EVOLUTION OF THE OPEN SYSTEM

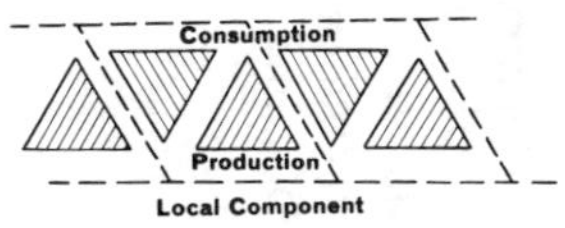

Most geographic areas are shaped both by consumption (Christaller's central-place system) and by production.

In any situation other than local self-sufficiency, the settlement pattern logically must be shaped as much by production as by consumption.

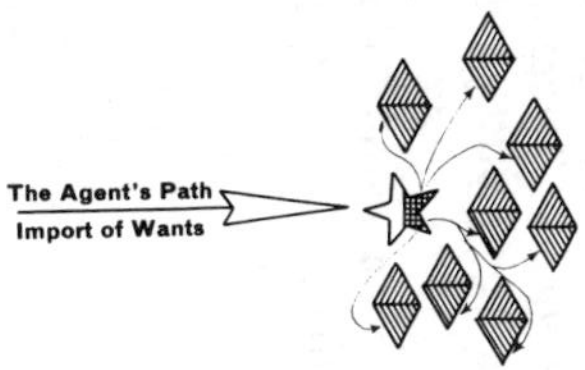

The creation of the agency of trade facilitates a geographically extended reflux of goods, often through the efforts of the same men who developed the distant market. The point of initiation of that market development is the probable spot at which a point of refluxive distribution will come into being.

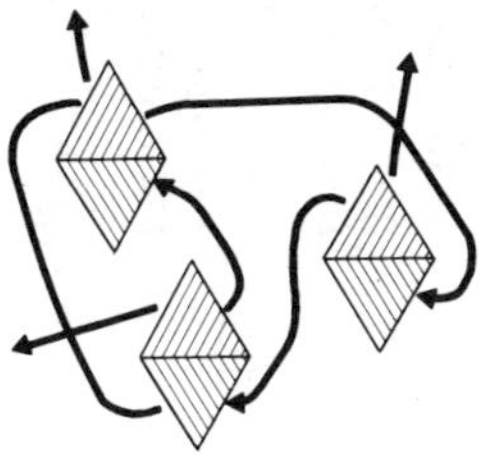

Thus, the interchange represented by trading comprises movements from one local component to another as well as within the component.

Furthermore, the notion of conflux on a trading place has two characteristics —the search for services and goods to satisfy wants and the search for a market for rural production. Given the consistency of rural produce over extensive areas, the probability is that the best markets will lie external to a cluster of local components.

In the market region for externally sold production there will be created similar points of initiation of trade, entrepôts, about which a similar open system of production-consumption local components will be ranged.

The agent's path can be of great extent, and may have intermediate points through which trade is organized and articulated.

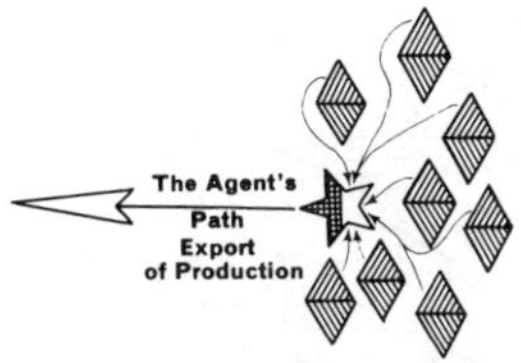

With the necessity to develop the external market there will be a strong force at work to establish trade-stimulating towns, points of initiation, from which agents will be sent forth to organize and conduct trade.

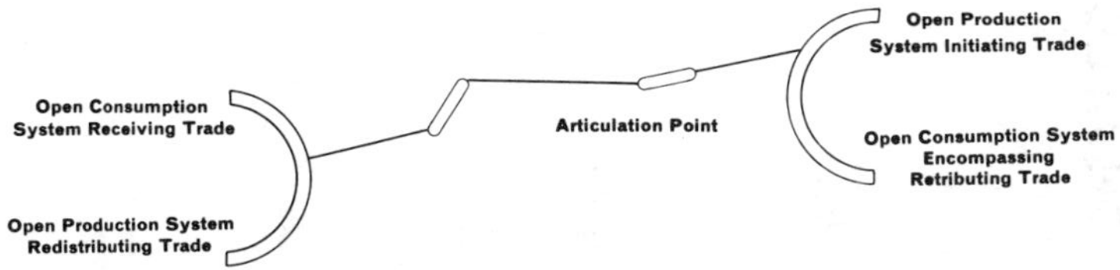

Figure 2

a moderately complex geometrical formulation of the sort of settlement pattern that might be envisaged as growing out of the movement of customers to purveyors of goods and services. As such it stood a paradigm for a special case, though one of great importance.[2] Although Christaller made no wider claim for his theory, time has seen his special case transformed by disciples into a general one—to the extent that a whole generation of geographical theorists has sought to account for almost any economic geographical pattern on the basis of central-place notions.[3] In that attempt there has been no regard for the strongly *endo*genic quality of the theory.

Before discussing this point, it is necessary to return to our conjectural history of trade. The next question to ask is, what happens when local self-sufficiency ceases to be total? Or perhaps we should say, what if areas never were totally adequate unto themselves? The simple answer is patent, that external trade will be introduced. Yet in seeking the origin of that trade we must take full account of the fundamental modification in the information structure that external trade would introduce. In the closed system of familial and local self-sufficiency [even in a classic central-place system, local self-sufficiency is essential to the central-place hierarchy (Fig. 1)], the direct association of producers-*cum*-sellers with customers assures that sufficient information will be interchanged to tell customers of potential products and sellers of potential markets. But with the advent of an external, open-system structure for trade (Fig. 2), which is envisaged as taking place when men realized that their local economy was either inadequate or materially impoverished, the information problem is introduced. First, we face the problem of explaining how men came to know of their local impoverishment, thus, how they came to realize the appeal of external trade. Secondly, we must account for the flow of information necessary to maintain an open system of such extent that sellers and buyers could not physically meet.

The solution to these two problems is to be found in the creation of a new component in the trading structure, that of the *agent of trade*. We may logically suppose that some group of persons, at some point in time, learns to produce goods in an abundance greater than the parochial need. The ability would father the desire to expand trade beyond the local area; but to secure that expansion it would be necessary (1) to familiarize the outside consumers with the abundance and its location, and (2) to familiarize the producer with the location of external scarcity in the good. In simple form, either the *pull* of scarcity or the *push* of abundance—or, most probably, the contemporaneous expression of both—creates a problem in information exchange. To forge the spatial tie

[2] Later I wish to suggest that the "special-case nature" of central-place theory attaches not merely to its explanation of retail trade, as opposed to all trade, but also to its particular application to settlement patterns derived from a feudal past (see Chapter 7).

[3] As evidence for the wide-ranging application of the central-place "explanation," reference to Berry and Pred's *Central Place Studies: A Bibliography of Theory and Application* (Regional Science Research Institute Bibliography Series, No. 1, 1961) should suffice.

someone must undertake as his basic job a geographical transfer of information. He must accept the role of agent of trade, standing as neither a producer nor a consumer directly. When we insert the agent (between these two trading poles in a geographical sense), we also introduce the fact of external shaping of the local conditions at either pole. The system is made *open* and subject to *exogenic change.* Under such conditions a number of the arguments of central-place theory become the explanation of a special rather than of a general case.

We have called this long-distance trade, but little explanation is required to substitute therefore the term wholesaling. If trade is to be carried on over great distances, or even moderate ones, an agent becomes essential, and, in the exercise of his office, there are obvious personal advantages gained by increasing the quantity of goods he handles. Later we shall see that a number of the structural qualities of long-distance trade make it necessary for an agent to deal in large quantities and to sell to retailers rather than to ultimate customers. For the moment it is adequate to note that once local self-sufficiency disappeared, wholesale trade was born. With that birth the determinate quality of economic geography was lost. No longer were neat assumptions as to how far and in what scale trade would flow possible. Even the total size of trade could not be predetermined, because no valid bounding line, short of the total world, was self-evident. Without the determinate qualities of retail trade, wholesaling was not amenable to analysis under any theory founded upon endogenic change.

The Absence of Work on the Location of Wholesaling

In such a pass, central-place theory fails to provide the system of analysis needed to account for the general structure and location of wholesale trade. In fact, the basic inability of Christaller's theory to handle trade practiced by agents probably accounts for the almost total absence of geographical literature dealing with wholesaling. Anyone seeking to work with such large-scale enterprise fairly quickly discovers that he is a pioneer in a vast and complex problem. Even the tools of analysis are different. In the study of retailing, geographers have been able to reproduce the nature of the trade through tallying customers and plotting their residence on maps, which leads directly to the reconstruction of lines of flow in the journey-to-shop. The surprisingly strict limitations on the ability of the individual family to consume tangible goods (inelasticity of demand) makes inference of scale in retail trade fairly easy. We have already noted that the assumptions of central-place theory allow some judgment of the total volume of trade expectable at a particular spot and the geographical reach of that center's appeal. But none of these measures apply to wholesaling. Trade dealings tend to be carried out in a most confidential way, and there is no real upper range of distance in the sale of goods. Furthermore confusion is created through the abstract quality of many transactions, wherein physical transfer of goods takes place along a different geographical linkage (between the manufacturer and the retailer or even the ultimate customer) from the line of the abstract paper transaction.

This mating of poor evidence and shaky theory seems a sufficient explanation for the insubstantial geographical work of wholesale trade in the past. Outside of geography the work has been similarly thin, though for different reasons. In the field of business administration the problem is definitional—a situation none of us finds hard to imagine when we seek to appraise what is meant by claims made by a vast array of businesses to sell at wholesale prices. Furthermore, students in economics and business administration are particularly vexed by problems of sorting out the integration of manufacturing, wholesaling, and retailing, for those activities almost irresistibly verge into one another today. Some argue wholesaling is unnecessary, others that it is really nonexistent, and still others that it certainly isn't important in our present concern for macroeconomics. With a few notable exceptions, these views tell us that there is little help to be secured outside geography in the search for an explanation of wholesaling location and the effect it has had on the broader geographical patterns of settlement.

To this end we must seek a new structural theory capable of dealing with conditions shaped by outside forces and compatible with the introduction of agency into the conduct of trade. That agency is continuing rather than periodic, as is demand satisfaction in retail trade. Thus, the play of experience and tradition is far greater in wholesaling than in its retail minion. The role of experience suggests that we should use a historical rather than a temporally abstract approach, and substitute induction for the rigid deduction of central-place theory. The search is for an order based on organized observation rather than on either abstraction or overly simplified knowledge.

Derivative Structural Theory in Geography

The sometimes shadowed world that is *man's* organized use of space assumes an increasingly important place in the thinking and research of geographers. Early in the history of the field our gaze rested almost exclusively on the physical pattern that nature provided for human use and on the tranformation man had brought to that landscape. The regional synthesis was the objective. We came to know the nature of the world through its parts, but not man's fundamental organization of that world for his own ends. Emphasis was placed on the parcelling out of space into units which were believed to possess individual self-sufficiency and coherence. Without some notion of order, however, it proved as difficult to understand the observed landscape as it is to explain the contour of a man's arm on aesthetic principles. All the field observation of an earnest and suffering generation produced no more than a catalogue of surface texture and blemishes.

Only when abstract reason was applied to observation were the sinews of spatial organization made easier to locate and explain. The approaches to geographical abstraction have tended to concentrate around either *mechanisms of organization* or *structural theory derived from observation.* In the first case a theory is advanced, such as that for the central-place hierarchy, on the basis of a single or a few timeless assumptions. The prime assumption in these mechanisms has always been that within spatial organization there is an evolution to a final state of order, with the concomitant notion that any departure from that state of order represents a clearly preliminary situation which will soon be "improved" so as to conform to the theoretical mechanism.

Derivative structural theory makes no basic assumption of a universal mechanism, but rather uses experience and history as the clay for a more plastic abstraction. Such an inductive approach allows for the comprehension of a greater range of spatial and temporal variation and the shaping of a less relativistic theory. Perfection and imperfection drop out, and evolution is viewed as both normal and uninterrupted. Finally, in the context of the present study, it is essential to note that the dynamic nature of derivative structural theory facilitates the analysis of problems in which it is impossible to disregard time and for which a geographically closed system cannot be assumed.

If economic geography is focussed on the sinews of productive activity, then wholesale trade is quite literally the bicep of exchange, connecting producer and consumer over considerable geographical distances. To carry the analogy to its logical conclusion, we might add that the length of the muscle is determined quite early in the history of an economic organism but use tends to give it strength and bulk. This use in itself points up the fact that wholesaling is a frequently repeated, easily institutionalized undertaking in which both ends of the sinew are normally quite determinate as to location, and both of fundamental importance in any location theory. The single polarity of central place if it ever can, in fact, be true, is meaningless in a study of wholesaling with its innately connective function.

The goal of this study is the understanding of present-day patterns of wholesale trade, but the path to that objective cannot be short. The truth is that wholesaling was the first means by which a local economy was increased in scale. When family and parochial self-sufficiency began to wane, wholesaling had to develop and the lines it took tended to become the pathways of history. Standard thought up to now, confined by central-place theory, has held that local areas focussed their search for goods *inward* on certain selected spots to the extent that in those places a threshold of demand could be reached, and that this focus, in turn, brought trading into existence. Here I hope to show that both history and logic belie any such initial gravitation of interest on "central places." Instead, trade began with the projection *outward* of demand from preexisting population concentrations, which were often *not* the direct outgrowth of retail trade. If we can accept the idea that settlement comes into being for reasons other than retail trade, then there is no innate validity in seeking to expand the particular and limited truth that people must cluster *inward* to buy goods into a universal mechanism of organization, as has been done with central-place theory. Instead, there is considerable logical appeal to the theory that an *outward* projection of demand, rather than an exterior surplus of production, led to the establishment of trade. In Chisholm's words we may emphasize that "Articles are made and services provided only because they are wanted or are thought to be wanted. Normally, it is demand that begets production and only rarely does availability of article create the demand therefor."[4] Only when demand focussed outward was there increased exterior production, the initiation of trade conflux on the city, and the establishment of a medium of exchange that would be necessary for trade. Without trade, there would be no basis for the satisfaction of exterior demands that leads to the creation of central marketplaces. Concentrating on the center alone hides the fact that some people fit into none of these retail systems except most infrequently. It is not adequate to explain merely where people shop without understanding also how frequently and why. In this context charting a path on the basis of single-pole organization, a central place would be foolhardy.

Initial Attempt to Apply Central-Place Theory

In the early stages of this study, the intent was to use central-place theory as the main frame for analysis. The available data for the pattern of wholesale trade in the United States were collected and amplified through detailed field mapping and appraisal of the local pattern of wholesaling in the San Francisco–Oakland–San Jose urban complex. An effort was then made to organize this information in terms of Christaller's concepts. The result was unnerving. Either much fact had to be rejected as errant, or much theory had to be fundamentally rewritten. In simple terms it became clear that central-place theory is most certainly a "special case," both of fact and of conception. As will be explained in the last chapter, where the relationship of the mercantile theory of settlement here advanced is related to central-place theory, Christaller's notion requires a closed economic and geographic system, and probably a history of feudalism as well, to secure much empirical support on the ground. When we look for an explanation of New World conditions and of trade in an open system, the south German paradigm is inadequate. Further, the experience gained in the decades since Christaller first wrote has made it manifest that any new theory must be more dynamic. Change in the geographical extent of a system could not continue to be overlooked in the context of the facts of wholesaling, which became, on collection, a powerful argument for the role of experience, commercial intelligence, and tradition in the shaping of patterns of economic geography.

Experience, as a Limiting Assumption, Substituted for the Region

The several approaches to geographic research outlined in the previous section call to mind one of the more difficult aspects of research

[4] Michael Chisholm, *Geography and Economics* (New York: Praeger, 1966), p. 147.

methodology that faces us. In a simplistic sense, geography is interested in all aspects of the world as a whole. But obviously this is an unattainable goal and we must adhere rigidly to the practice of introducing *limiting assumptions.* In the work of Edwardian geographers the intellectual stricture was that of the *region,* a division of the indigestible whole that was thought to have validity and to be separable from other regions by defensible and observable bounding zones. A half-century of methodological wandering, which would have tired even Parsifal, has come no closer to the regional grail than a glimmer just over the edge of the last field study area. Rather reluctantly most geographers have abandoned the regional concept. In its place they have put a new set of limiting assumptions drawn deductively in the work of social scientists. Von Thünen, Cooley, Haig, McKenzie, Hoyt, Christaller, and Lösch have all provided physical designs from which geographical order could be reasoned and explained. Certainly the most significant of these in the work of geographers was Christaller's central-place theory, but we have already noted its failure as a conception in the geography of wholesaling.

The search for a formulation that might help us to understand the locational order within wholesale trade shares the need for limiting assumptions. A theory of wholesaling, with its strong dependence upon schedule and tradition and on the establishment of ties that persist in time and expand over distance, demands two qualities that are not found in central-place theory: (*1*) an ability to deal with support and change induced from outside, and (*2*) a dynamic quality, for the long and persistent ties found in wholesaling give evidence that history strongly influenced the shaping of trade patterns. This is a call not for historical geography, which seems largely to shun any strictures other than parochial and to lack structural organization other than chronological, but rather a call for the use of history in the elucidation and understanding of a *theoretical and limiting* formulation. In other words, we need theory based on *experience* with that experience acting as a seine to separate the significant detail from the irrelevant fact—*a structure derived from an ordered appraisal of historical evidence.* We do not seek to tell where the King of Spain went, but rather why he was on the road and what the journey availed. This is historical economic geography rather than economic historical geography, with its emphasis on the past for its own sake.

The World Shaped to the Geography of Wholesaling Is Our World

The inward focus of both regional and central-place theory has concentrated the attention of geographers on the dependent nature of wholesale trade rather than on its influence on regional development. The fundamental assumption has been that wholesaling takes a role given to it within a strongly hierarchical structure, making its entrance only when the settlement fabric becomes dense and complex. Yet history tells us that wholesaling and cities came to America on the Mayflower and were just as much tools of pioneering as the axe and the hoe. This use of the city as a tool of mercantile conquest can be followed back in history at least as far as the Phoenecian "plantations" such as Cadiz, founded around 1100 B.C. The Venetians, in turn, became rich and powerful by using the Fourth Crusade as a fact-gathering and entrepôt-founding expedition to Phoenicia itself. In the commercial conquest of America only the battleground was new, the objective and the system were ancient. Strangely, the deductive search for limiting assumptions has never perceived either that mercantile purpose or its settlement weapons. Only by thinking in systemic terms can we create the dynamic theory which in turn illuminates the truth that economic geography works in an open system as well as a closed one. We should not forget that it was a Yankee merchant ship which first brought America to the attention of a parochial world, making wholesale trade a fundamental national occupation and the American diet and material culture the most geographically complex examples on earth. Thus trade did not grow *out* of American economic development, rather it induced that development.

The view that trade spreads outward from the city, or the mother country, leading to the complex flowering of the rural economy at home as well as in a distant plantation of staple-producing colonists seems an accurate summation of the economic system under which most of the new lands developed. No doubt the characteristics are found at least in outline even in the emerging economies of African states. The point that is essential to our argument is that these are *initially* trading economies rather than producing economies in the narrow sense of local self-sufficient systems. Settlement takes place and populations expand in fairly direct proportion to the avenues of trade open to the colonists, unlike the situation typical of the old lands of Europe and Asia where population and settlement reflect political ends and general health conditions.

Movement over distance is manifest in any trading economy and on a scale that soon requires the creation of a specialized occupation to handle those shipments. Once such an intermediary steps between the producer and the ultimate consumer, the *localized* economy is no longer an organic necessity and the requirements of the producer alone do not predetermine the site of trade. Central-place theory assumes that all trading acts are a unity as to time and place but the truth is far more varied. A good may be bought or sold through another man's agency, which brings about a division of the act of selling and buying into several discrete parts. Geographically such a division is of high significance as it permits a range of choice as to the location of trade: the mechanism of central-place theory cannot account for such nondeterministic variation because that theory cannot encompass several possible resolutions of the location problem, with the choice of solution determined by history rather than by geometry. In place of Christaller's simplistic explanation we must create a theoretical structure that is anchored to the city and whose external characteristics can be varied with nature and stage of economic development. The specific application to wholesaling of this argument is found in the simple fact that the sinews of the fundamental structural system of Western economic life are largely, though not wholly, those of wholesale trade. If central-place theory is specific to retail trade and services and culturally relative to lands that suffered the fractionating setback of the feudal era, then the external-influence notions here presented are specific to wholesale trade and the mercantile- and industrial-capitalist eras which grew up so rapidly from the sixteenth through the nineteenth centuries. These, after all, are the times when the world we know was shaped, and this is the workshop whose organization we understand and still, for the most part, use.

The Establishment of Derivative Structure

In the basic pursuit of a location theory of wholesale trade, this study must accomplish several objectives. The definitional problem is dealt with in the next two chapters. With some understanding of what this trade is from the geographer's viewpoint, we return to our fundamental purpose of establishing a derivative structural theory to explain the location of the activity itself. Chapter 4 sets forth on the search for an understanding not of wholesaling itself, but rather of what it accomplishes in the geographical economy of the world. This quest is both conceptual and historical, seeking first the functional structure of the trade and then its development before the onset of the commercial revolution induced by the discovery of the new lands and the route to the Indies. That revolution was so critical to wholesale trade (in that it opened an empty field to the merchant's efforts), that it seems desirable to treat it separately, as is done in Chapter 5. Further in that chapter it is possible to bring down to the present the quantified appraisal of the location of wholesaling in the United States by looking at the period for which mass data are available. This, in any thorough study, is too absurdly short a time—1929–1963—to serve as the basis for a serious attempt at understanding wholesaling. Thus, *we must unite recent statistics with longer historical experience to shape adequate location theory.* Chapter 6 seeks to apply some of the ideas on wholesaling location advanced for large areal systems to the local system of the metropolitan city. The final chapter sets out to accomplish two things, a formal statement of a theory for wholesale location—and its relationship to Christaller's central-place theory—and a proposal for a parallel economic-settlement model to help fill the gap left by the "special case" nature of Christaller's work. This Mercantile Model of Settlement allows us to close in on the most essential aspect of wholesale trade, its outward-looking quality, which has always been a missile of cultural and material advance and a powerful force in shaping both settlement and the geography of economic activity.

TIME IN TRADE. Before setting out on this search it is necessary to enter one explanatory note as to terminology employed. Already frequent mention has been made of the importance of time in the study of wholesaling and in the pages that follow, time becomes almost an organizing principle in this appraisal. This fact grows out of the significance that temporal values have in the location and success of wholesaling, essentially in the very creation of the trade. Without going into detail here, it may be pointed out that the various components of the spatial structure of wholesaling—the hearths where trade originates; the trade structures along which it flows, both initially and in established pattern; the entrepôts which introduce trade to an undeveloped area; and the much more advanced stockroom wholesaling districts—all are concepts of time as well as space. It is well at this early juncture to come to appreciate that time is as much a geographical factor as the range of a good or its threshold of provision. And that time factor shows up in the concept of "demand accumulation" which is discussed in Chapter Four.

GEOGRAPHICAL THEORY DRAWN FROM NEW LANDS. In the study of trade there is logic in looking for its geographical pattern in those places most strongly and simply shaped by the activity. Europe for much of its history was more characterized by impediments to trade—by dynastic and ethnic preserves forbidden to outsiders and by enclosing fear which walled men up in strongpoints little related to anything but physical survival—than by any powerful reach of trade. As Henri Pirenne has proposed, it was the return of trade that closed the Dark Ages and began to break down the parochial barriers that made Europe feudal. It is not important to our purpose to assess whether trade remained a trickle during those fractionated times: certainly there can be no doubt that the geographical organization of settlement and economic activity under feudalism did not grow from the mercantile considerations that here concern us.

Turning to America, the aboriginal story reads the same, with a brutal division of space among defensive groups. Quick change came, however, with the total and rapid destruction of that narrow frame when the European merchants sailed onto the shore in the sixteenth and seventeenth centuries. Their vigorous takeover erased the savage slate to the extent it became easy for trade to shape massive patterns of settlement related to those of commerce. In the absence of detailed survivals of records of business from colonial time, we may use the observable settlement pattern as a substitute source of information. The quick and complete re-creation as Anglo-French-America makes a strong argument for the use of a "new land" in attempting to establish theories of trade and settlement relationship free from the too heavy political tradition of a feudal land.

When asked in 1929 to look at the application of general work in economics to such a new land as Canada, Harold Innis stated his belief that "A new country presents certain definite problems which appear to be more or less insoluble from the standpoint of the application of economic theory as worked out in the older highly industrialized countries."[5] The failure comes from more than just the contrast between

[5] Harold A. Innis, *Essays in Canadian Economic History,* ed. Mary Q. Innis. (Toronto: University of Toronto Press, 1956), p. 3.

staple-producing and industrial countries; it grows out of the fact that economic theory is shaped, by its objectives, to be uniformitarian in spatial concepts. Implicitly, save in notions of population or market potential, space has no meaning beyond that of a cost factor. To "simplify" thinking, theory tends to obscure the geographical and historical influence and put forward a rigid economic determinism which envisages the present as the norm. Such assumptions may have utility for analyzing economic systems, but in the study of locational patterns these strictures are myopic. Turning to the simple analysis of structure here proposed, it is obvious that no high level of abstraction is attempted. Abstraction is deliberately rejected to avoid the destruction of time and space contrasts, whose understanding is the central purpose of this study. We seek here not to know the economics of wholesaling but rather to begin to understand its history and present geography.

If we might exclude that past and reason that the needs of the present determine the geography of wholesaling, far more sophisticated statistical procedures might be used. But the prime significance in mass trade of the early start, the established line of trade, and defensive and aggressive trading policies—all ideas which can be fully developed only in a context of historical economic geography—argue for the study of structure rather than of a timeless mechanism. Here our study seeks not to classify cities into structural compartments but rather to ferret out the forces that have shaped the broad features of wholesale trade. The ontogeny presented is intended as a limiting abstraction, an historical–geographical template on which to measure time and functional position for wholesaling. As data of any detail for wholesale trade began to be collected only in 1930, the use of a statistical abstraction in place of the ontogeny is patently impossible. Rejecting the notion that the present is the norm, as well as the catechism, a device for relating diverse and spotty evidence gathered over a long time period had to be designed. The methodology has been shaped to explain, with the same concepts, the location decisions of the Venetians in 1199, the Massachusetts Bay Company in 1629, the Crane Company in the 1920's, and wholesale grocers in the 1960's. No apology is made for failing to use elaborate statistical procedures. In a study of certain present-day problems they have great value but in an attempt to put the vast historical geography of wholesaling in a rational context, statistical analysis is too restricted in time and too rigid, unreal, and deterministic.

CHAPTER 2 *the nature of wholesaling*

Often things are so self-evident to intelligent people that definition becomes pedantry. When, however, we speak of wholesale trade and its geographical conditions, the chance for a common understanding, even among intelligent men, is small. There may be agreement that wholesaling is different from retailing, but that distinction merely removes the extremes of doubt without telling what wholesaling is. As is normally the case, the argument lies in the borderland, and to add to the confusion, many merchants who know better call themselves wholesalers in order to borrow a cachet of cheapness. Some manufacturers carry on wholesaling functions, and some retailers have activities that would seem to be wholesaling but are not so called. Thus, the innately transitional nature of wholesaling combines with the highly ambiguous use of the term by most people to make certain that definition is far from pedantry.

The Link Between Wholesaler and Retailer

Since it is impossible to use self-designation for classifying wholesalers, we may ask what, by general agreement, does the process of wholesaling accomplish? The most straightforward reply would be that wholesaling connects, in trade, numerous producers of commodities with an even greater array of consumers. Direct access to the customer by the producer is impossible both because of the great numbers involved *and* because of the considerable geographical distance that often separates the groups. There must, nonetheless, be ties between the two groups that will (1) provide the producer both with a market and with some idea of its demands and their scale, and (2) assure to the consumer access to products of a determined type at a specific time. Once the localized structure of trade was modified, such links could not be secured without a wholesale *agency* whatever it might be called. The important consideration is not who bridges the gap—a span built from either shore will cross—but rather that the tie is made. One of the critical changes in economic geography came when three elements of trade appeared—producer, agent, and consumer—where before there had been but two. With the interposition of an agent the local economy became part of a much broader regional economy, and it should not be overlooked that this expansion of horizons led to pioneering far more readily than did retail trade.

When man ceased to provide for all his own needs, a whole series of consequences resulted. Dependence on others introduced trade but along with it *control* through exchange in a regulated market or fair within the settlement. In a relatively isolated society, trade does not necessarily imply transportation other than in its most restricted sense of men moving about a self-contained settlement. The farmer may bring his produce in from the field and the townsman must carry that produce home to his house; but in an economic or geographical sense, transportation hardly exists. Yet there is the possibility of strong retail trade.

What this tells us is that when there are many producers and many consumers retail trade will come into being; and so long as places of production and consumption are identical, wholesale trade need not be resorted to: *Only when there is geographical separation of production and ultimate sale need wholesaling intervene.* The main qualification of that statement would come in the case of complex economies wherein the number of producers of a particular type of product would be great enough to make the potential customer anxious to be able to compare, in one place, the product of the lot. Then a special trader would set up shop serving both as retailer and as wholesaler. For example, a grain merchant might both be a traveling buyer at wholesale setting a purchase price and a retailer of part of his purchases. The fact that the early organization of trade was neither precise nor totally consistent need not disturb us; but *it is safe to say that retail trade began with the end of familial self-sufficiency, and wholesale trade began with the end of local self-sufficiency.*

The role that the trade plays, linking customers and producers over distance, makes wholesaling of major importance in the explanation of patterns of economic activity. Until men came to enjoy personal mobility, retail trade had to be pretty much a local affair with the market town standing alone in the retailing structure. The close spacing of those towns, until the railroad, served mainly to show the low personal mobility rather than some geometrical determinism. Above the market towns there were emporia, but they were, as we shall see, mainly wholesaling structures such as the Stourbridge, Champagne, and Leipzig fairs. In simple fact before George Stephenson built the first railroad between Stockton and Darlington in 1825, the merchant had to seek out his customers in local markets, save for a minute aristocracy whose peregrinations were, in large measure, socially rather than economically derived. Thus for all but a handful of people, personal shopping had to be extremely local, carried on either at stalls in their village marketplace or from the peddler's bag in their own kitchens. Any sort of hierarchy of trading journeys only had meaning for the aristocracy. In such a context the notion of the range of a good is socially limited, and mainly of significance in an aristocratic rather than a democratic framework.

After 1825 there was a rapid increase in ease of movement. Whereas the aristocracy had mobility before the tracks were laid down because they could engage in periodic residence moving seasonally from the city to the country, the middle classes were less footloose. In the 1780s the pioneer cotton merchant Jedidiah Strutt, although a very wealthy man, had to send his wife as agent to London because he could not leave his works in Derbyshire.[1] The trains made periodic absences possible for his entrepreneurial group. As a result, an elaboration of classes of retail towns could be introduced with enlargement of the geographically mobile classes, and the local markets began that slow decline which has never been reversed.

For wholesaling these events are indirectly important. As the hierarchy of retailing expands, the geographical structure of wholesale trade must obviously be transformed, as surely as the length of the anchor chain must reflect the depth of the water. When most men had little opportunity to leave the close confines of their neighborhoods, the wholesaler had to go forward toward the customer to reach markets close to the farms and the factories where the early mill workers were employed. But when men could travel to larger towns for shopping, the wholesaler had less far to go. *Given a static population, the function of the wholesaler will be carried farthest forward along the chain of supply in the earliest stages of trade.*

Where Will the Store Be?

In retail trade the essential contribution of the merchant is that of holding a stock of goods relatively close to the prospective customers. The phrase "relatively close" is, of course, one of variable meaning depending upon the interaction of a number of factors, the most important of which in this context is transportation technology. In the Middle Ages "relatively close" meant, as Dickinson has shown,[2] at a point where the country man could go to market and return during a single day. Even in the American Middle West such a view persisted, as evidenced by the close spacing of potential market towns under the Congressional land survey.[3]

[1] R. S. Fitton and A. P. Wadsworth, *The Strutts and the Arkwrights, 1758–1830, A Study of the Early Factory System.* (Manchester, England: Manchester University Press, 1958), pp. 25–26.

[2] Robert E. Dickinson, "The Distribution and Functions of the Smaller Urban Settlements of East Anglia," *Geography,* 27 (1932): 19–31 and "The Markets and Market Areas of East Anglia," *Economic Geography,* 10 (1934): 172–82.

[3] William D. Pattison, *The Beginnings of the American Rectangular Land Survey System, 1784–1800,* University of Chicago, Department of Geography, Research Paper no. 50 (1957).

Given a static population, either through high mortality or constant emigration, we should expect to find a change in the location of retail activity with any improvement of individual mobility, as we noted above. Once the possibility of concentration was at hand, the economies to be secured through tapping a larger market could be compelling in the location of shops. What this says is that Christaller's concept of threshold is not immutable, and we cannot logically talk of a threshold without taking into account the time and stage in which it is found. With increasing personal mobility the size of the retail threshold will grow, and the location of the retail store will move from the consumer toward the producer (so long as population remains static), and the length of the link from wholesaler to retailer will be shortened.

If we now assume an increasing density of population within a region, we introduce a force that works opposite to the impact of transportation technology. As an example, we might take the contrast between the countryside of New England and of the "Upper Midwest." In New England there has been a growth in rural, albeit rural-nonfarm, population for a generation now. This growth has been contemporaneous with the concentration of retailing in larger towns and units of selling. But in New England there has not been the destruction of small-town shopping simply because the increasing density of population has counteracted the concentrating tendencies. In this case the threshold may be said to have enlarged, as Christaller would have it, but the location of retail trade has not shifted. Given the situation in northern Wisconsin and the Upper Peninsula of Michigan, where population has been declining,[4] the enlargement of a threshold means automatically the shift of trade from smaller to larger central places. The result will again be that of shortening the wholesale–retail link that had been forged in the previous generation of trade.

When we assume a dynamic nature for the ties between the retailing function of the town and dispersed customers in its trade hinterland, we may also assume that changes in the wholesale–retail linkage are possible. If the location of shops shifts toward larger towns, that link is shortened in a situation wherein the wholesaler remains stationary. But the movement of shops and shopping into fewer and larger centers may also facilitate the shift of wholesale activity "back along the pipeline" toward the place of manufacture, securing to the wholesaler economies of scale to match those enjoyed by the retailer with an expanded market threshold. Thus, we must view the wholesale–retail link as elastic in nature permitting locational shift in the warehousing activity as well.

The rural storekeeper of the past had several advantages over the country merchant today. To begin with, he had almost complete command of his market because consumers could not seek alternative sources of supply. This meant that fewer people were needed to secure a marginal support. In addition, the demand for goods on the part of

[4] John R. Borchert, *The Urbanization of the Upper Midwest: 1930–1960,* Urban Report no. 2 (February 1963) *Upper Midwest Economic Study,* pp. 13–18.

the local residents was much more automatically to be focused on the nearest general store; and the restricted market thus developed in any one place limited the diversity of goods that could be traded. In short, the early American storekeeper, for example, had an isolated market which allowed him to limit product diversity at any particular time and yet command the total economic value of that market. With a reduction in his control over any of these three market characteristics—*exclusivity, demand accumulation,* and *product diversity*—he would be forced to seek a larger threshold. The larger market brought in by changes in transportation is geographically derived, whereas the preserved threshold in southern New England today is demographically based.

Geographical Sheltering of Trade and Demand Accumulation

The geographical sheltering of the trading act must be clearly understood if we are to work out the location theory for *all* trade (as opposed to Christaller's work on retailing and service trades alone). If the local market is protected, it can be dealt with as a "closed system," at least at any one time. Goods on hand are sold to the local population, in relation to principles that have come to be called the threshold and range of a good. So long as the customer *lacks the mobility to shop elsewhere, and insists on immediate satisfaction of his demand,* it is hard to fault this central-place model. Once we open up either of those restrictions, the neatness disappears and a more inclusive theory must be devised. In place of the timeless closed system we must create one that is either periodically or continuously "open" to the outside world.

To see how that opening up works, we may view the extension in both time and space aspects.

It seems probable that not even the trading system in the Middle Ages was fully-closed, if we think of the system as one made up of both time and space relationships. The local customer was restricted to the use of the local merchant when he asked immediate demand satisfaction. However, when he exercised customer patience, he could have a wider choice because, from early times indeed, peddlers and chapmen had met just such unusual demands, even if they could do so only once or twice a year. Thus, it is historically and logically inaccurate to build an analytical model totally on such a closed-system as Christaller did.

A geographically circumscribed view of a city's trading area was never fully valid, but it became even less imaginable when a wide range of men gained more personal mobility, through the spread of railroads. At that pass many men were able to look outside the local trading system, if they had either an unusually pressing demand (extreme time utility) or a demand for an uncommon type of good that could wait long enough to permit satisfaction at a distant trading place, as an adjunct of other travel (extreme place utility). Such adjunctive demand satisfaction is an observable attribute of travel—if it were not so, it is doubtful that retail merchants would be so generous to convention bureaux—and today we may even hold that for many persons there is no precise economic appraisal of the acceptable length of the journey-to-shop. If the act of shopping is the test, rather than its cost, the long drive may be the end in itself.

In light of these expressions of demand accumulation, which open up Christaller's closed system, the customary *division* between the location theory of retailing and wholesaling loses a great amount of reality. Viewed the other way 'round, the use of an open system in the analysis of retail as well as wholesale trade is also indicated.

Wholesaling and Locational Equilibrium

Before going on, we must consider Lösch's theories on economic location patterns. Lösch was primarily interested in the "equilibrium" of location patterns, or more correctly, in states of conditional equilibrium. In Lösch's view there are five conditions of equilibrium, or forces that account for the location of economic activity. In this manner, the conditions of equilibrium of economic locations stand in the place of physical forces of equilibrium in physical location. As he notes, "Nature works according to laws, but man acts according to his *idea* of laws. In other words, nature *must,* man *may,* act correctly. In order to do so he must have some conception of how he shall act. As to economic equilibrium this means that *in order to guide his activities* he needs insight into the conditions of equilibrium." [5] These are:

Condition 1: *The location for an individual* (trading establishment) *must be as advantageous as possible.*
Condition 2: *The locations* (for trading establishments) *must be so numerous that the entire space is occupied,*
Condition 3: *Abnormal profits* (for trading establishments) *must disappear,*
Condition 4: *The areas of supply, production, and sales must be as small as possible,*
Condition 5: *At the boundaries of economic areas it must be a matter of indifference to which of two neighboring locations they belong.*[6]

By integrating Christaller's ideas with the notions of equilibrium put forward by Lösch, we have tools with which to assess changes in the hold on an area's trade that a town may have at different times. Under conditions of effective isolation from competitive trade centers, the meaning of the threshold is different from that in conditions of mobility and competition. In isolation trade is completely controlled by the actions of a single merchant or a group of merchants at one point, who can thereby concert their actions with greater ease. With no opportunity for effective geographical competition, the level of profit can be increased, allowing what in Löschian terms would be conditionally

[5] August Lösch, *The Economics of Location,* trans. William H. Woglam (New Haven: Yale University Press, 1954), p. 93.

[6] Ibid., pp. 94–97. The italics are in the original, which here has been amended parenthetically to read specifically for wholesale and retail trade rather than for all economic activity as Lösch intended.

abnormal profits. With such high profits the threshold population might be lowered, since each consumer would pay more for the item he buys than he would in a state of competitive trade. What this says is that in pioneering economies, wherein geographical isolation can be maintained, trade will commence with a population much lower than that required to maintain activity once transportation is established among competing trade centers. Because central-place theory was developed in terms of an old and highly integrated economic region (south Germany), this modification was not apparent to Christaller. But in the study of wholesaling the point is critical, because we shall find that the *stage of economic development* assumes far greater importance in accounting for wholesale location than it does in retail location.

As pioneer economies grow, two things might happen. Given a stable threshold value, we would expect new trading centers to come into being allowing Lösch's second and fourth conditions to obtain; the entire space will be more effectively occupied the greater the number of trade centers it contains, and additional centers in a defined region will obviously reduce the size of the individual areas. With an increase in the support available in the region, it would also be possible to envisage a reduction in the "abnormal profits" available to the pioneer merchant. In that case there might be no additional establishments. The expression of more support would come in the pressure that would be put on the individual merchant to lower his prices or risk competition. The possibility of a static trading pattern with lower unit profits is critical to our understanding of wholesaling, since it says, in effect, that with an enlarged market, the place of trade can remain the same if other conditions of business change (e.g., if the individual entrepreneur discourages competition by reducing "abnormal profits").

Such a defensive tactic is not open to the retailer to the same degree that it is available to the wholesaler. The retailer depends upon clear, direct, and rapid access to his customers. Except in certain trades, the cut-rate store a day's journey away is not, in normal terms, competitive. Thus, Lösch's fourth condition calling for the smallest possible tributary areas relates directly to retail trade. In wholesaling the merchant is not so dependent upon clear, direct, and rapid access to his customers, mainly because his relation to them is different. In the wholesale–retail linkage the trade is repeating and formalized. Just as the retailer sticks to a particular line of trade, he also tends to maintain *particular wholesale linkages,* both in type of goods sought and in sources of supply. The conduct of his business is facilitated by reducing the supply problem to a routine. To do that, the establishment of regular linkages is of great value.

The creation of consistent ties between retailers and the wholesalers who supply them leads to a more formal structure than that relating consumers to the shops where they make their purchases. *By institutionalizing the line of trade, custom replaces chance, and it is possible to maintain ties over a considerable distance without the risk of competitive intrusion.* The quite formal nature of relations between retailers and wholesalers is to a degree the result of contractual arrangements. Wholesalers traditionally have been sources of credit for retailers. They also have aided local merchants in securing scarce commodities and thereby commanded their loyalty. And we should not forget that the retailer forges his own chains when he maintains customary wholesale ties; the goods he handles tend to have consistent qualities that his customers come to expect a particular product to have. To change wholesale supplier might well mean to change the character of his goods and thereby injure his retail trade.

The institutionalization of wholesale–retail linkages allows the wholesaler to manipulate Lösch's conditions of equilibrium to his own advantage. He is able to assert the fundamental importance of the first condition (the location for the individual trading establishment must be as advantageous as possible), and at the same time argue that small tributary areas are not necessarily of advantage to his retailer customers. Wholesaling by its nature is not automatically facilitated by numerous locations, nor is it automatically more efficient when small. And the ability the retailer has to shift agency to a different wholesaler if he is convinced that the former ties meant excess costs gives the notion of excess profit a new meaning. In a situation similar to the conditions of frontier shopkeeping, the retailer may well argue that as a wholesaler increases the size of his market, he is able to exist on a smaller mark-up, even accepting a static threshold. Thus, so long as the retailer is free to change his line of supply if the wholesaler takes abnormal profits, the maintenance of existing trade ties may allow the retailer to gain more price flexibility than he would secure by shifting patronage in an excessively divided market. If the wholesaler lowers his price with an increasing market, the retailer can either lower his price and hope to enlarge his clientele, or maintain former levels and earn higher unit profits. This choice is demonstrated in the United States where the small shopkeeper, particularly in a relatively isolated location, is able to mark up his goods more than his urban peer because in the country proximity to customers [condition 4: areas must be as small as possible] is advantageous and competition is lacking. Competition, access to stores, and customer satisfaction in retailing would be maximized in some cases by encouraging larger wholesale firms in fewer large centers, while permitting a proliferation of retailing outlets. The main restriction on this relationship is the difference in transport cost of bulk, as opposed to small, shipments.

There Is a Difference Between Wholesaling and Retailing

It should now be clear that wholesaling has location characteristics that differ from those of retailing or manufacturing; but so far theoretical constructs seeking to explain the location of economic activity have related only to production and retailing, not to wholesaling. Alfred Weber provided an analytical system that has been valuable in explaining where industry would locate,[7] and Christaller has provided one similarly for retail trade and services. Only Lösch, and then only in a

[7] Alfred Weber, *Theory of the Location of Industries,* trans. Carl J. Friedrich (Chicago: University of Chicago Press, 1929).

very general frame, has been concerned with large-scale distribution. Thus, understanding the location of wholesale trade and the fundamental forces that shape it, requires the reformulation and enlargement of existing location theory.

A Definition of Wholesaling

Up to this point it has sufficed to refer to wholesaling as connecting the producer of commodities with the retailer. Even this limited definition, however, is more precise than is common. Merriam-Webster's *New International Dictionary* for years identified wholesaling as the "sale of goods by the piece or in large quantity," [8] which in addition to a seeming universality, also displays appalling imprecision. The *Shorter Oxford Dictionary* is less universal but no more precise when it defines wholesale in the adjectival sense as, "In large quantities, in gross. In a large way, in large numbers or amount, in abundance," and notes that the term was first used in English in 1601 in the descriptive sense and in 1645 to mean selling a commodity by wholesale. No doubt much of the trouble in giving to wholesaling any logical bounds comes from these largely literary uses of the word.

In the business administration literature and the United States census the definitional emphasis shifts from scale to method of operations. Whereas in common literary language the term wholesale denotes large-scale, when the word is employed, in business practice, through institutionalized usage, certain connotations are given, which must be taken into account in any study of wholesale trade as an economic undertaking. To understand that institutional view we must aknowledge that the use of *goods traded* is the key. If the purchaser uses the goods in pursuit of his trade, rather than simply to satisfy personal or family needs, the transaction is normally considered as wholesaling, whether actual resale takes place or not. The true meaning of this definition is best shown in a brief discussion of the business administration literature.

The first full-scale investigation of wholesaling appeared in a book by Theodore Beckman and Nathanael Engle originally published in 1937.[9] Prior to that time there had been a sparse literature concerned mainly with the entrepreneurial aspects of the trade. Beckman and Engle believed that the first study of wholesaling in this country was that issued by the Harvard Business School in 1916 centered on "Operating Expenses in the Wholesale Grocery Business." [10]

Out of the study of wholesaling by men engaged in business administration grew an awareness of the distinct quality of the trade. Beck-

[8] This definition appeared in the second edition of Merriam-Webster's unabridged dictionary. The third edition is far more precise and up to date, terming a wholesaler "a merchant middleman who sells chiefly to retailers, other merchants, or industrial, institutional, and commercial users mainly for resale or business use—called also a *jobber*."

[9] Theodore N. Beckman and Nathanael H. Engle, *Wholesaling, Principles and Practice* (New York: Ronald, 1937). [This book was revised in 1939 and subsequently reprinted a number of times.]

[10] Ibid., p. 5.

man and Engle proposed three basic criteria to distinguish wholesaling from retailing.

In formulating a scientifically correct or theoretically sound definition of a wholesale transaction, consideration must be given to three basic factors or criteria, including:

1. Status or motive of the purchaser, i.e., the customer and his purpose in making the purchase.
2. The quantity of goods involved in the transaction.
3. Method of operation of the concern.[11]

It should be obvious that wholesaling involves the sales of one entrepreneur to another, intended for resale by the second. Resale in this case needs to be defined very broadly to allow the inclusion of the commodity wholesaled in another fabrication or in the conduct of the purchaser's business. Thus, a restaurant may buy a half-dozen bottles of Worcestershire sauce for use by its customers in exactly the same manner those customers would use the sauce at their own table. This would be a wholesale transaction, not because of the number of bottles purchased at one time, but rather because the supplying of restaurants is carried on by wholesalers rather than retailers. The restaurant's consumption of a single commodity might not greatly exceed that of a family strongly given to the use of that product, but the total consumption of food by the restaurant would be such that wholesalers would find business relations with it profitable.

In these terms classification is determined by the nature of the establishment's customers: the wholesaler would have customers who were entrepreneurs rather than simple consumers. The entrepreneur would anticipate securing a profit from the investment he had made in the wholesaler's product. Thus, sales to restaurants, hotels and other catering institutions, to manufacturers, public utilities, government agencies, and nonprofit institutions, to farmers for farm use, to builders and contractors, to men independently carrying on a profession or service, and to retailers, would equally qualify as wholesale trade. *The purpose rather than the scale of purchase is the determinant in these instances.*

As Beckman and Engle note, "Contrary to the layman's conception, actual practice during this century has vitiated quantity as a basis for characterizing a transaction as wholesale or retail."[12] In the garment industry, for example, sales at wholesale may be of very small size, frequently repeated, and based entirely on the wholesaler's rather than the manufacturer's maintaining stocks. Under such conditions it is unrealistic to pretend that the unit of purchase provides any clear criterion for separating wholesaling from retailing.

[11] Ibid., p. 17. Thus only in part is wholesaling a matter of [large] scale. Certain operations include very large-scale purchases—anthracite trade sales averaged a number of tons each and yet were mainly retail transactions—so more than size of order had to be considered.

[12] Ibid., p. 20.

The Accumulation of Demand

Nevertheless, scale is not an insignificant definitional criterion. A wholesaler extends terms and wholesale prices to such small-unit purchasers as, for example, those in the garment trade, because the total of purchases over the years that he may anticipate warrants wholesale trading. However much the family may consume, unless it has an entrepreneurial function, its totality of purchase must be less. This tells us that in the division of retailing from wholesaling it is crucial to introduce the dynamic concept of *accumulated demand* in place of unit demand. Discussion of the establishment location aspects of accumulated demand must be put off to Chapter 3 but here we may attempt a simple definition of the concept. By accumulated demand is meant the total demand for a good by one purchaser, either individual or institutional, over a relatively long period of time. Normally that time would be no less than a year, and in some cases it might be considerably longer. Whatever the period of time, there is the implication that the demand is relatively constant in expression but still susceptible to periodic satisfaction. Such demands as those for perishable foods, fuels, shelter, and other "necessities of life" cannot normally be periodically satisfied and thus are not characterized by accumulation of demand.

In cases where it is not possible to accumulate demand for periodic satisfaction, it is probable that measures of scale rather than motive of purchase will be the determinants of a wholesale status for the transaction.

Beckman and Engle round off their criteria for determining which transactions should be classed as wholesale by arguing that the method of operation of the establishment or firm must be taken into account. In part the argument is narrowly related to business practice—in such matters as discounts and pricing policies—while in part it is of considerable value in assessing location requirements.

Normally, a retail establishment is characterized by the following:

1. it is open to the general consuming public;
2. *sales are made over the counter;*
3. usually in small quantities; [but]
4. in large numbers [of individual transactions];
5. to ultimate consumers who buy for direct consumption; [and]
6. at so-called retail prices that are not subject to various types of discounts.[13]

That retail trade takes place over the counter, is crucial to the notion of a central place. Conversely, since wholesaling does not depend upon over-the-counter sales, it is able to disregard some central tendencies, and substitute therefore formally and intricately structured linkages with its customers.

The definition of wholesaling that emerged from investigation, during the thirties, of the *modus operandi* of wholesalers was essentially

[13] Ibid., pp. 21–22. Italics supplied.

purposive. If a transaction was undertaken to allow the purchaser to profit from its use or sale, or in such quantity that no reasonable question of individual use could exist, then it was wholesaling.[14]

On such a background, the description of wholesale trade in the *Standard Industrial Classification Manual* of 1957 is easily foreseen.

> This division (wholesale trade) includes establishments or places of business primarily engaged in selling merchandise to retailers; to industrial, commercial, institutional, or professional users; or to other wholesalers; or acting as agents in buying merchandise for or selling merchandise to such persons or companies.[15]

In many ways the geographer might wish that the scalar element had loomed larger in this more careful analysis of wholesaling. When wholesaling is regarded as related to warehousing, a tradition still maintained in British English through the virtual synonymity of the two words, there is little ambiguity about the physical expression of wholesaling. Trading on a large scale required massive buildings, and those buildings would have a characteristic urban location, which simplified the geographer's work of explaining wholesaling locations. Unfortunately, the complex nature of trade means that no simple and obvious scalar sorting takes place. As we shall point out later, the "industrial park," so fashionable in real-estate circles today, is as much the wholesaler's home as is the dilapidating zone of discard in the wake of the central business district.[16]

Types of Activities Subsumed in Wholesaling

In a trade characterized by three hundred thousand establishments in the United States [17] the probability of consistency is small. Within such a total are firms that may share purposes with the group but have a

14 Beckman and Engle defined wholesaling and retailing basically as opposites. "Wholesaling includes all marketing transactions in which the purchaser is actuated solely by a profit or business motive in making the purchase and, if the goods are bought from a concern operating substantially as a retail establishment and such goods are not intended for resale, the quantity is materially in excess of that which might reasonably be purchased by an ultimate consumer.

"Retailing includes all marketing transactions in which the purchaser is actuated solely by a desire to satisfy his own personal wants or those of his family or friends through the personal use of the commodity or service purchased; it also includes transactions involving the purchase of goods for industrial consumption, from a concern operating substantially as a retail establishment, in quantity not materially larger than that which might reasonably be purchased by an ultimate consumer." Ibid., p. 23, italics removed.

15 *Standard Industrial Classification Manual* (Washington, D.C.: U.S. Department of Commerce, Bureau of the Budget, 1957), p. 147.

16 R. E. Murphy, J. E. Vance, Jr., and B. J. Epstein, "Internal Structure of the Central Business District," *Economic Geography*, 31, No. 1 (January 1955): 21–46.

17 Data used here will generally be drawn from the *Census of Business* (Washington, D.C.: U.S. Department of Commerce, Bureau of the Census, 1963) in its various parts, which will be cited for identification by census report number and censal year, thus: Table 1, Wholesale Trade, United States Summary, BC63-WAI (1963), which may be read simply as (B)usiness (C)ensus of 19(63), report (W)holesaling (A)rea Series, number (1). The total number of wholesaling establishments in the United States in 1963 was 308,177.

distinctive manner of carrying on their functions. For that reason students of wholesaling have come to distinguish among wholesaling enterprises not by size, because size loses meaning as we have seen, not by purpose, because all wholesale merchants seek to tie the manufacturer with the penultimate purchaser, but by the type of business organization. For our purposes such an organizational classification has weaknesses, but by distilling them out, we may secure a palatable and useful product.

At first it may seem that the definition of wholesaling as that trade that sets out to supply retail and institutional purchasers with goods that they may sell or use to produce a profit, furnishes us with a net of sufficient discrimination. Such it is in the sense of distinguishing between wholesalers and retailers, but not in yielding differences among merchants of similar business organization and location requirements. In a study of geographical location such as this, we need to divide our preliminary catch in terms of normal habitat so we shall know where again to look for them.

The fundamental connective quality of the trade is already to be seen in Adam Smith's classic description of wholesaling. In describing the four ways that capital might be used, he tells us one is "in transporting either the rude or manufactured produce from places where they abound to those where they are wanted"; this is the manner in which the "capitals" of all wholesale merchants are employed.[18] Classic as this may be, it leaves unanswered the questions where the merchant will set himself up in business and how he would secure and distribute the goods in which he is trading. In a trade characterized by its connective function, the wholesaler might well locate either near the source of the supplies in which he deals or close to the largest group of his customers; he might restrict his interests to a narrow segment of trade but exercise his efforts over a wide territory, or try to cope with a broad array of demands in a restricted local area.

Types of Wholesaling

Depending upon the observer, there are two, three, six, or more different *types* of wholesaling. Types in this sense means specifically the manner in which the merchant conducts his business and the relation he bears to sources of his supply and locations and organization of his purchasers.

THE MERCHANT WHOLESALER. The first wholesalers almost certainly were *merchants* in the original sense of that word, men who negotiated or trafficked in goods on a large scale, particularly in foreign trade. These men set out to trade as a livelihood and seldom said nay if profit could be secured. The Boston goldsmith, John Hull, was such a merchant who used his earnings at his trade to invest in ships whose cargoes he then offered for sale in Boston or elsewhere. We find Hull telling his captain, "if you goe to liverpool I suppose [you will load] course manchesters and tapes, etc. Bremogen Knives [Brummagem, that is, Birming-

[18] Adam Smith, *Wealth of Nations,* bk. II, chap. 5.

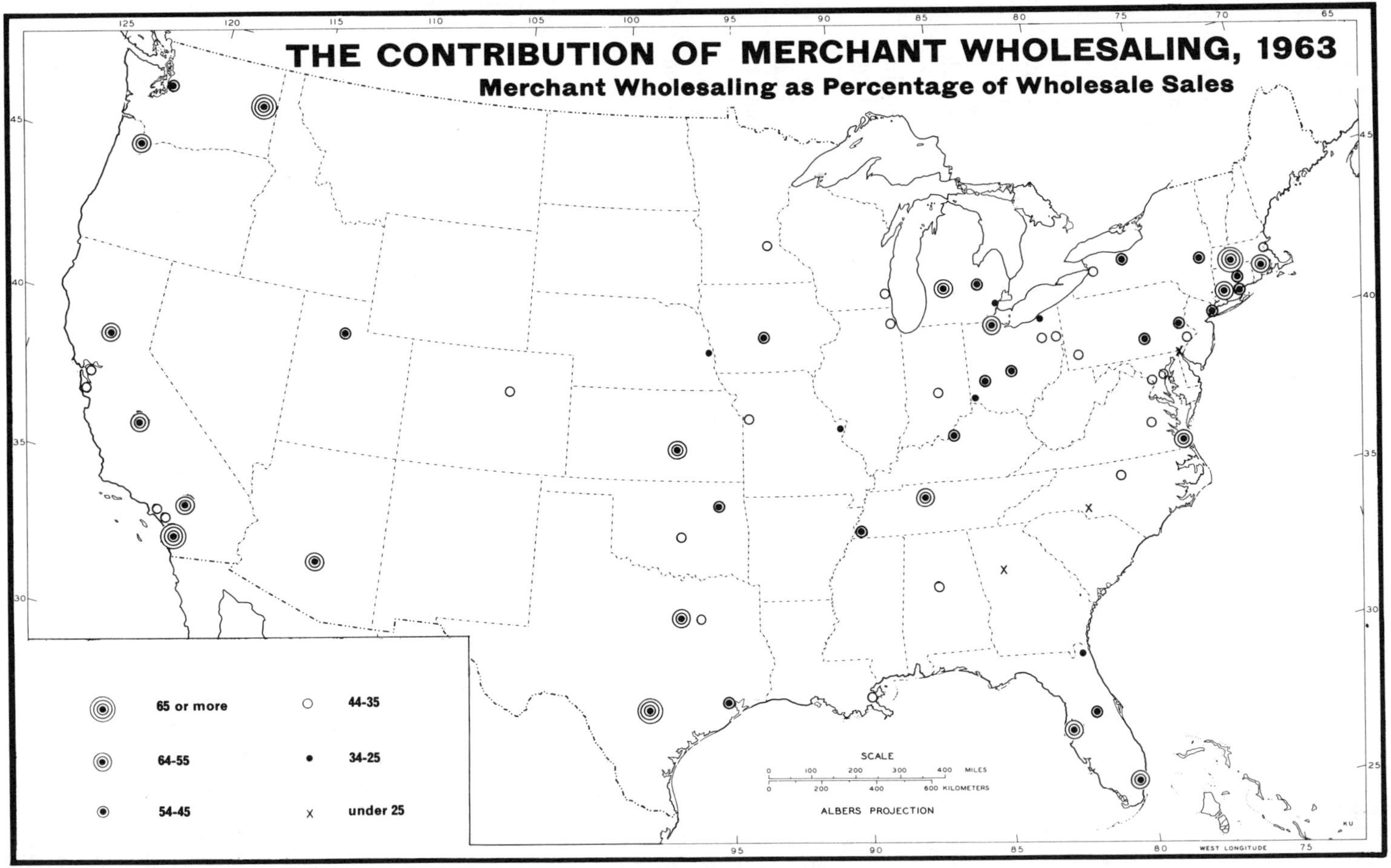
THE CONTRIBUTION OF MERCHANT WHOLESALING, 1963
Merchant Wholesaling as Percentage of Wholesale Sales
65 or more
64-55
54-45
44-35
34-25
under 25
SCALE
0 100 200 300 400 MILES
0 200 400 600 KILOMETERS
ALBERS PROJECTION
WEST LONGITUDE
KU

Figure 3

ham goods] etc. Red Cotton duffals, such things as are made att or neere the place but may be bought cheape." These when arrived in Boston might be sold at shipside to the local people or carried farther to New York or Virginia in Hull's ships.[19]

From that day we have called the class of standard entrepreneurial trader on a large scale a *merchant wholesaler.* First as individuals and later as corporations these traders formed the fundamental body of wholesalers. They are "establishments usually known as wholesalers, distributors, jobbers (except those in the clothing trade . . .), foreign trade merchants, limited-function wholesalers, etc.—distributors primarily engaged in buying, taking title to and, where customary, physically storing and handling goods made by others and selling the goods at wholesale principally to retailers or to industrial, institutional, and commercial users."[20] Over the years the merchant wholesalers have been the largest component in the trade, and it is this type of establishment that matches most the standard conception of wholesaling. In the most recent census of wholesale firms and establishments, just over two out of three were classed as merchant wholesalers.

THE MANUFACTURERS' AGENT. In contrast with merchant wholesalers are *manufacturers' sales branches and offices.* "These establishments differ from merchant wholesalers in that they are owned by manufacturers or mining companies and maintained apart from manufacturing plants, primarily for selling or marketing their products at wholesale. . . . Manufacturers' sales branches are distinguished from sales offices by the fact that the former maintain facilities for the psychical storage, handling, and delivering of at least a portion of the merchandise sold."[21] In the constant effort of industrialists to secure the noncompetitive introduction of their products into a specific market, the sales branch offers advantages. The firm is able to "push" its own wares exclusively, skirting the efforts of producers of similar products, except when those producers also engage in exclusive branch operations. Thus, once the distribution system is introduced, all seem quickly to feel that they must adopt it, and defense can be had only by recourse to identical protection. The use of sales branches is widespread in certain industries—tires, electronics, pharmaceuticals, business machines, for example—wherein all important producers may be forced to care for their own wholesaling. Since in recent years many of the items thus traded are those with a rapidly expanding demand, it is not surprising that the sales volume has expanded far more rapidly than the actual count of places of business.

Manufacturers' sales agencies and branches also differ from merchant

[19] Samuel Eliot Morison gives us a splendid description of John Hull in his *Builders of the Bay Colony* (Boston: Houghton Mifflin, 1930), pp. 133–82, wherein he shows that broad objective and nearly universal accomplishment were attributes as much of the seventeenth- and eighteenth-century merchants as of their learned contemporaries. (The quotation is from p. 171.)

[20] *U.S. Census of Business:* 1954, vol. 4, *Wholesale Trade-Area Statistics* (Washington, D.C.: U.S. Department of Commerce, Bureau of the Census, 1956), Appendix A, p. 3.

[21] Ibid., p. APP–4.

wholesalers in that their managers have little or no purchasing function. The source of goods is determinate and unchanging, or else the agency would have little to offer the producer, and only in the sales of goods is the branch operation flexible. There can be no doubt that these representatives of the manufacturer along the road of distribution *are located in line with an over-all strategy of the firm, but in this case a manufacturing firm,* in its attempt to maximize its sales or earnings. In this way the location of sales agencies is hard to tie to simple dynamics. The branch may not, in fact, take its location from the analysis of any assumed hierarchical system of trading towns, but rather from policies of the firm to "invade" a particular market or to provide a certain level of service to gain volume of sales, with the idea of shifting the earnings expectation from the wholesaling function to the manufacturing level by using wholesaling largely to secure increased sales and enhanced economies of manufacturing scale.

The fraternity of manufacturers' sales agencies and branches has two sections, that in which the establishments receive, store, and distribute the goods in which they trade, and that in which the trade is limited to the securing and remitting of orders for goods. Where goods are physically handled, the sales agency differs only in quality from the establishment of a merchant wholesaler, in the sense that it acts in the manner of the latter firm but within the narrow range of one manufacturer's goods. But a sales agency dealing only in orders rather than goods has relatively little kinship with the merchant wholesaler and is likely to seek quite different premises. In a situation where goods are not dealt with physically, space and transportation lose importance in favor of *communication* and obtrusiveness. To be near potential customers and to be able to collect and convey their desires to the parental manufacturing firm is the purpose of the establishment. This means that centrality becomes more significant than it would be to the wholesaler with stocks, and the ability to subdivide the wholesaling function into a number of relatively small dispersed units encourages the parent firm to create a number of sales agencies (without stocks) in reflection of the threshold of trading activity *they are willing to accept.*

THE BROKER. These sales agencies are extensions of the manufacturing corporation. When they serve many rather than a single master, a line of distinction is drawn, but with small contrast, and the establishment is that of a *broker* or sales agent. If he works simply as a sales agent, not exercising judgment and discrimination in the selection of goods for sale, the difference from manufacturer's agencies is minor and of no interest to us save for a single consideration; that is, that the independent sales agency must earn its own living rather than depending upon corporate policy, and often subsidy from manufacturing profit, for existence. In other words, unlike the manufacturers' branch, there is no one to pick up the bill for the independent sales agent. Thus, the independent selling agent may have contractual ties with a number of manufacturers, serving them as a captive salesman, which limits or excludes any purchasing function. Nonetheless, such an agent must locate himself so that he can secure an economic return for his activities. This leads him to set up shop more in the fashion of a merchant wholesaler than would his counterpart in the industrial sales office.

The sales agent is normally a man dealing in orders rather than goods. For that activity an office generally suffices. Such is also the case with *brokers,* whom we might define as men or groups of men whose function it is to maintain a "market" in a particular product. They collect information from retailers or other potential customers. They thus must have close association with their customers, and they must also be able to transform that information into purchasing actions directed at competing producers. The broker must exercise judgment far more constantly and literally than his sales agency associate. Perhaps if anyone is the heir to the medieval merchant, it is the modern broker who attempts to provide a service and nothing more. In knowing what goods will be needed sometime in the future and placing anticipatory orders, in judging where that desire for goods will be largest or strongest, and in directing the flow in that direction, does the broker earn his hire. The modern sales agent is to the broker as the New York horsecar driver of 1824 was to the scouts in the fur trade in the Rocky Mountains. Both are honest trades, but only one is exciting.

Brokers then must be in the market. The essence of their contribution is in helping to set prices for goods. Auction companies represent the simplest form of market operation wherein the wholesaler undertakes to do no more than provide a place for trading in a particular commodity. Title rests with the producer until the auction sale is made, when it vests with the purchaser, thus leaving the auction house to make its profit entirely on fees charged for providing a place in which a market price could be determined.

Other traders who help to fill the ranks of agents and brokers, as a classification, are called commission merchants. Their function is to serve producers by selling their product in a particular market.

The Factors. Almost limitless variations of the function exist. In the sale of raw cotton such commission agents often are called factors, and yet there is a class of bankers who stand as textile factors in the sense that they undertake to finance the production for a particular order by advancing to the manufacturer the cost of staple for its production. It is necessary to distinguish between these two—the trader in the commodity and the financier of production. Each carries a title derived from the early stages of merchant trading when factors were men who organized production by domestic workers. Part of that organization was akin to what we find the wholesaler doing today in anticipating demands and setting orders for their satisfaction at some time before the demand itself is openly expressed.

Another part of the medieval factor's job was that of providing the material from which to carry on production. Today such factoring is a financial rather than a physical undertaking, and quite apart from the *appraisal and satisfaction of future demands* which is so much the wholesaler's purpose. Textile factors then fall outside our range of interest, but their collateral relatives, the cotton factors, do not, because they carry out the functions of anticipation that we expect of wholesalers.

EXPORT-IMPORT AGENTS. Export and import agents fall into the general class of agents and brokers. These men dealing in foreign trade may have characteristic methods of operation, but they work fundamentally at providing a linkage between producers and consumers. Their knowledge of market conditions and sources of supply is for hire, and it earns their income. In some cases they may undertake certain financing functions, but in general the importer or exporter is a trader more than a banker. Often he may not take physical possession of goods; but it is his orders that start the flow of goods and determine particularly their origin. Just as in the case of commission merchants and auction companies, the import or export agent secures advantage from association with his kind.

One of the central and timeless problems of trade is the setting of fair prices, and it is this close association of men engaged in a specific trade that allows prices to be set. For that reason, wholesale trade, in the classes mentioned, shows a degree of flocculation quite apart from any possible dynamics of central place. City size, or even manufacturing activity, is not a sure key to the location of the center of trade (as opposed to the retailing town or center of manufacture). The persistence of the wool and leather trade in Boston long after the secular decline of the New England woolen and shoe industries is tied to this particular force. In somewhat abstracted trading, wherein ownership rather than physical possession is transferred, the critical concern is not any sort of Weberian configuration, but rather the setting of a fair price. Once the medieval economic system with administered prices declined and the universal church was split, only the market could accomplish such pricing, and the market required a *center of trade*, which became our commodity market with its population of agents, brokers, commission merchants, and foreign traders.[22]

The closing paragraphs of any classification are likely to be rather broadly inclusive and arbitrarily neat because there usually are individual cases that do not fit. Wholesaling is normal in this respect. Several of the commonly identified types of establishment are tidied up for the occasion. The classification has so far centered on the business organization of the wholesaling establishment and its place in the pro-

22 Alfred Weber in *Theory of the Location of Industries*, trans. Carl J. Friedrich (Chicago: University of Chicago Press, 1929), presents an analytical system that is thought to account partially for the location of manufacturing by weighing the impact of the flow of raw materials. The system, however, is impotent to account for the location of trading in raw materials, since those transactions are "weightless" in the Weberian thinking; thus there is no way to set up a location configuration for commodity trading. We are fortunate that history is less synecdochical and allows us to consider raw materials in several aspects, one of which is ownership distinct from physical presence. The setting of a price takes place among owners rather than among warehousemen, so the location problem for a study of wholesaling is different from, and far more comprehensive than, that of the flow of raw materials.

ducer–consumer line of flow. But the class of *petroleum bulk plants* and *terminals* has no virtue beyond one of convenience and synthetic consistency. Some such terminals are the undertaking of oil producers, others are independent operations, whereas a few are tied to consumers alone. This is a class apart from the logic of the classification so far, but it is the class used by the Bureau of the Census; and since it stands locationally distinct, only consistency would be served by rejecting it.

The final class of wholesaling commonly distinguished is that of *assemblers of farm products,* which stands as arbitrarily neat. Included are collectors in the simple sense—grain elevators—collectors who stand rather in *locum tenens* for the farmer—short-period livestock feeders and marketers—and outright processers—processers of vegetable products to the extent of elaborate packing operations. The consistency among these establishments is that they all stand at the head of any pipeline between the farm and the table. As such they may transform the product, they may keep it in marketable shape, or they may merely forward it. We shall have relatively little to do in this survey with the collection of agricultural products, so to accept the classification as it stands is convenient and not of great moment.

CHAPTER 3 the american pattern discerned

The five broad categories of wholesale trade discussed in Chapter 2 contain all establishments that operate substantially as intermediate level traders. If we may judge from shifts in membership of classes that have taken place since the original statement of the basic classification here used, i.e., between 1929 and 1963, there have been a number of minor reassignments that show a degree of restlessness.

The first nationwide assessment of wholesale trade resulted from the initial attempt of the United States government in the matter of business censuses. Earlier tallies had dealt with "occupations" or "trades" rather than with businesses as such. Thus, a worker would be classed according to his statement as to the work he did, but few men would normally call themselves wholesale workers, and none did in the early counts. Thus, we can find men in trade, but we are not told what aspect of trade they followed. During the period of rapid population growth and economic expansion after the Civil War there was a tendency, shared by all developing countries, to undervalue the contribution of trade and exaggerate those of agriculture and manufacturing. In what might be called the first modern census in the United States—and one of the earliest anywhere—that in Massachusetts in 1875, scant attention is given to trade even though trade stood at the head of the list of occupations in the number of employees in most of the cities, if not the towns.[1] We learn that by then there were 68,000 workers in trade, a

[1] See, for example, the Table of Selected Occupations for municipalities given in *Census of Massachusetts, 1875*, vol. 1, *Population and Social Statistics*, pp. 479–571. In the same census a full volume (vol. 2) is given to manufacturing, presenting 935 pages of data. In the *Census of Massachusetts, 1885*, no greater attention was shown to trade at the same time that the manufacturing data volume (vol. 2) had grown to 1,543 pages.

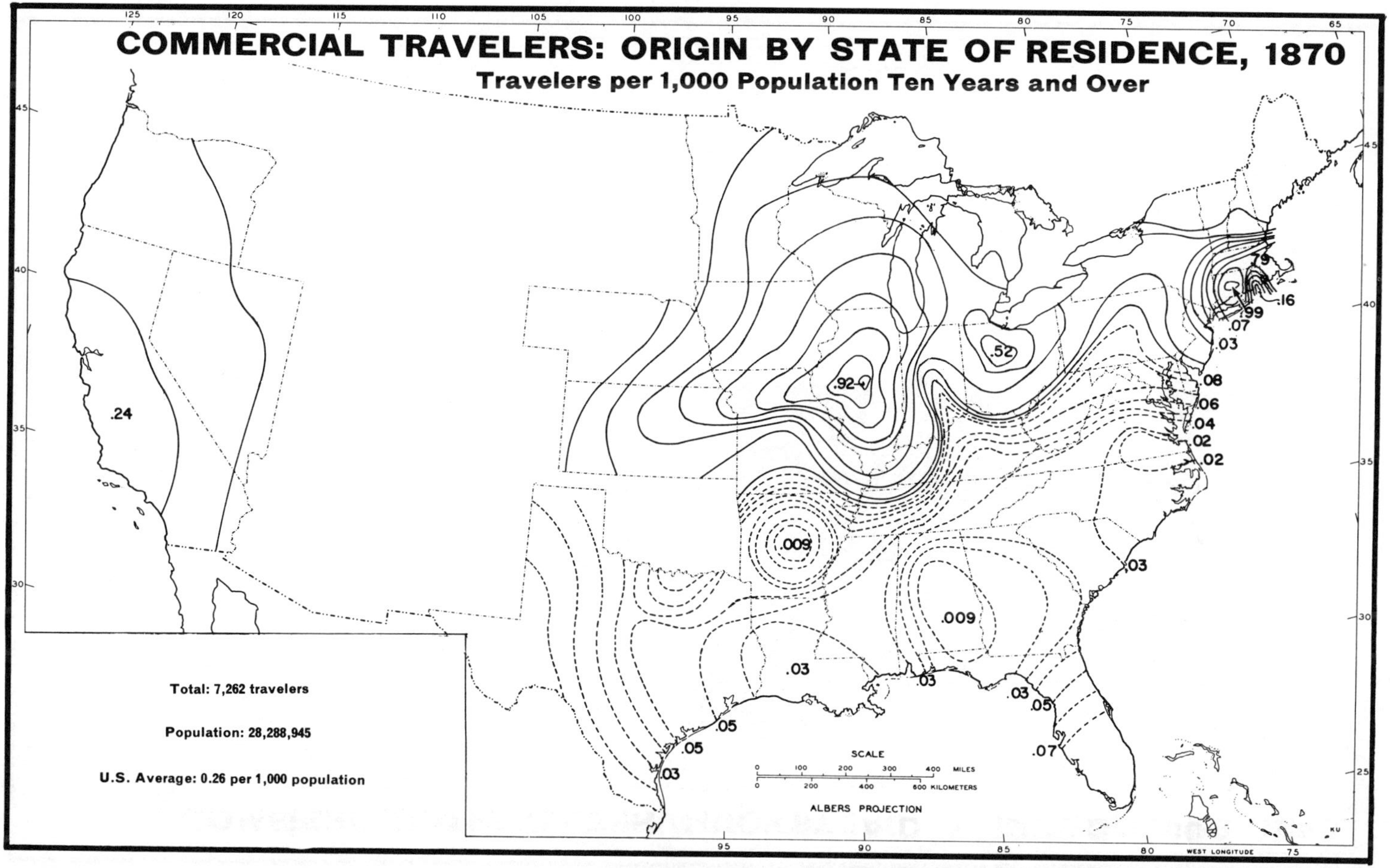
COMMERCIAL TRAVELERS: ORIGIN BY STATE OF RESIDENCE, 1870
Travelers per 1,000 Population Ten Years and Over
Total: 7,262 travelers
Population: 28,288,945
U.S. Average: 0.26 per 1,000 population
.24
.92
.52
.79
.16
.99
.07
.03
.08
.06
.04
.02
.02
.009
.009
.03
.03
.03
.05
.05
.03
.05
.07
SCALE
0 100 200 300 400 MILES
0 200 400 600 KILOMETERS
ALBERS PROJECTION
WEST LONGITUDE

Figure 4

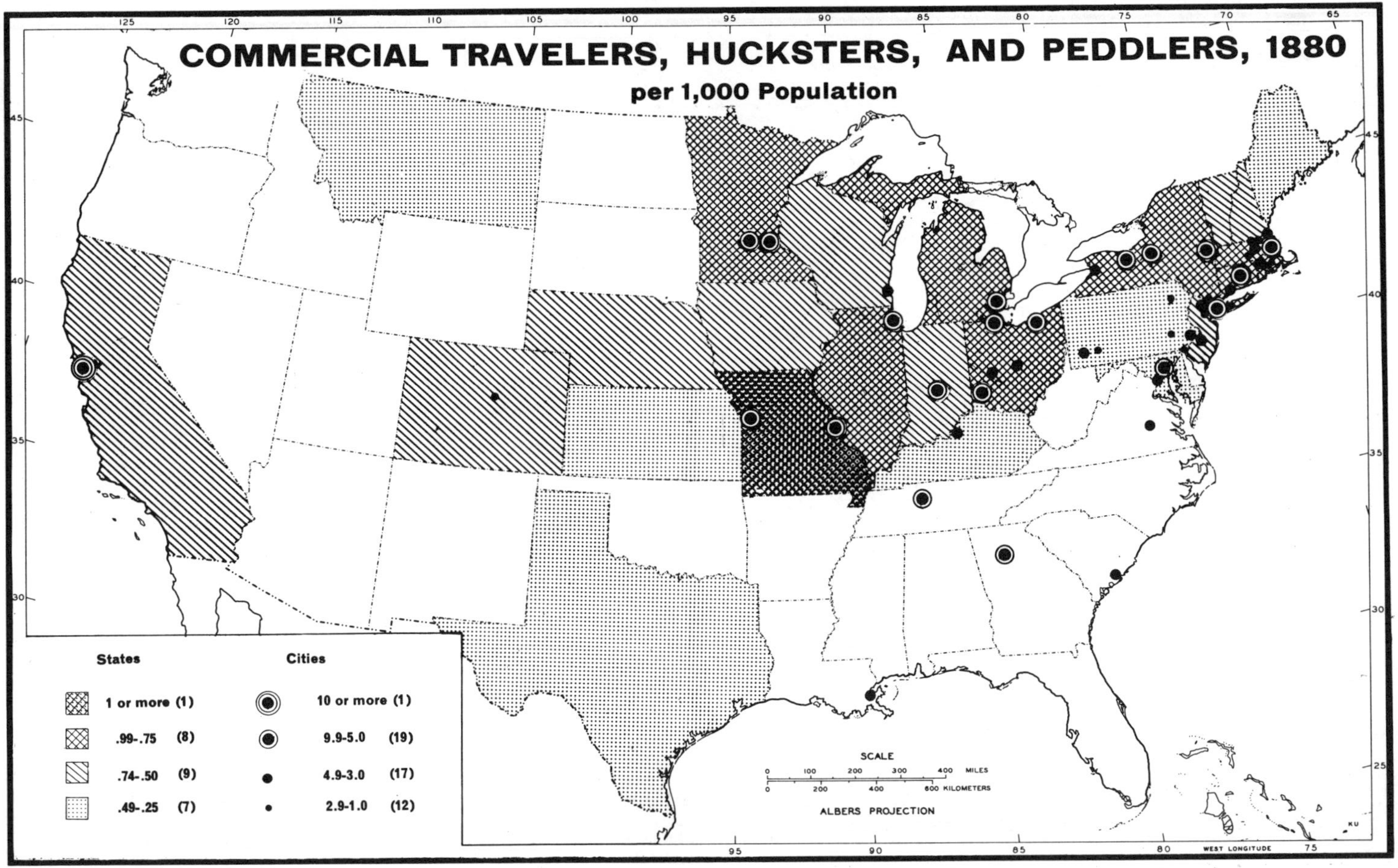

COMMERCIAL TRAVELERS, HUCKSTERS, AND PEDDLERS, 1880
per 1,000 Population
States
1 or more (1)
.99-.75 (8)
.74-.50 (9)
.49-.25 (7)
Cities
10 or more (1)
9.9-5.0 (19)
4.9-3.0 (17)
2.9-1.0 (12)
SCALE
0 100 200 300 400 MILES
0 200 400 600 KILOMETERS
ALBERS PROJECTION
WEST LONGITUDE

Figure 5

group as large as that in agriculture.[2] Admittedly Massachusetts would be one of the first states to find trade on a par with farming, but the fact that such a tertiary activity could bulk so large by 1875 tells us that the absence of *interest* in trade at that time was the result of short-sighted notions of relative economic importance rather than objective importance.

THE FIRST WHOLESALE CENSUS: 1929. Unfortunately, although we may tell how many workers in trade there were at the close of the first post-Civil War decade, we cannot sort out the groups in retail trade from those in wholesaling. Only with the creation of the first Census of Distribution in 1930 was information available. That tally recorded all wholesale establishments in operation at the beginning of 1930 and gave the nature of their operations during the calendar year of 1929. Thus, the census is normally attributed to 1929 and is sometimes referred to as the 1929 Census of Distribution, the wholesaling part. It was, however, an integral part of the fifteenth Decennial U.S. Census appearing as Volume II of the Census of Distribution.[3] That undertaking established the general character of all subsequent censuses of wholesaling, and changes over the years have been relatively small. In the 1929 material there is obviously a degree of experimentation in the classes of trade. Since no such census had previously been taken and a consistent classification of business types was two decades in the future, with the Standard Industrial Classification of Nonmanufacturing Industries first proposed in 1940,[4] the 1930 census had to try out various classification schemes in which to record the masses of data collected for the first time. In this light the enumerators' efforts were remarkably successful. The classification had seven classes of wholesaling establishment:

1. The largest group comprised the firms called *Wholesale Only,* the determining characteristic for inclusion therein being the nature of the business organization. These firms, whether "service" and "limited function" wholesaler, exporter, importer, converter, or supply and machinery distributor, were "*engaged in the buying or selling of merchandise on their own account at wholesale.*" In 1930 these single-minded establishments made up 41.81 per cent of all wholesaling premises.
2. In contrast in terms of business organization were the *Manufacturer-Owned Outlets,* which were usually tied to a specific manufacturer to the extent that the purchasing function became atrophied and the selling function greatly emphasized. The everyday operations of *Manufacturers' Sales Branches* (9.73 per cent) might not differ greatly from those carried on by "wholesalers only" but the range of commodities sold, like their provenance, was strikingly different. Along with a minor subgroup, that of *District and*

[2] Computed from the table Occupations by Name, *Census of Massachusetts, 1875,* vol. 1, pp. 573–609.

[3] *Fifteenth Decennial Census, 1930,* Census of Distribution, vol. 2, *Wholesale Distribution* (Washington, D.C.: U.S. Department of Commerce, Bureau of the Census, 1933).

[4] *Standard Industrial Classification Manual* (Washington, D.C.: U.S. Department of Commerce, Bureau of the Budget, 1957), p. 1.

General Sales Offices, wherein supervision of selling rather than selling itself was the objective, these producer-tied establishments bulked far larger in sales (23.29 per cent) than in numbers (9.83 per cent).

3. *Chain-Store Warehouses* stand essentially as the reciprocal of the manufacturers' branches in that they represent the backward extension of the retailer along the producer–consumer linkage, whereas the sales branch is the forward thrust of the manufacturer. In 1930 these retailers' warehouses accounted for a slight 0.83 per cent of all establishments, but they conducted a business eight times overweighted (2.79 per cent of all sales).
4. Petroleum products' *Bulk Tank Stations* display a universalism unusual to this classification by organization. Some stations were wholesaling branches of oil producers, others were basically independent merchant wholesalers, whereas still others were tied to chains of filling stations. The test in this case was not organization but rather the narrow function undertaken by the establishment. Within that firmly drawn boundary, *Bulk Tank Stations* formed a group rather in proportion accounting for 3.22 per cent of all wholesale establishments and 3.03 per cent of all sales.
5. Under a definition of wholesaling as a collection as well as a parceling activity, *Assemblers and Country Buyers* of farm products march in the column. The individuality of these buyers is great, ranging from elevator operators to lobster pound shippers. All, however deal in large masses of goods and have major functions in conferring place utility on those commodities and helping to establish prices. These rural collectors made up 20.12 per cent of wholesalers in 1930 and yet conducted only 6.68 per cent of the sales. The impact of a geographically dispersed production pattern thus enforced its own pattern of dispersion through the proliferation of buyers needed to purchase its product.
6. As noted before, the class *Agents and Brokers* is broad and diverse, but it is relatively consistent in organization. These men may buy and sell for others, they may merely buy and sell as agents, or they may buy and sell on their own account. In all those actions they serve largely by collecting and using information to carry out their office of assuring a functioning and rational "market." It is knowledge and experience that the agent or broker sells more than any particular commodity that give the volume of sales a somewhat abstract quality. However that may be, the total of agents and brokers is 10.84 per cent of all wholesale establishments, and their net sales amount to 20.58 per cent of total sales.
7. As in all classifications, there were those that did not fit with ease in any of these categories. For them a portmanteau "all other" class was included, representing only 0.29 per cent of all establishments and 0.98 per cent of sales.[5]

CENSUSES IN THE THIRTIES. The next toting up of wholesaling grew out of the economic collapse that began just at the close of the initial census. In 1933 a special Census of American Business was carried out with only minor changes in categories.[6] Wholesalers Only became Whole-

[5] The definition of these classes are abstracted from "Definitions of Types of Wholesale Establishments" in 1930 *Census of Distribution,* vol. 2, *Wholesale Distribution,* pp. 43–48 and the data from chart I of the same volume (p. 2).

[6] This special census material is not as readily available as are decennial censuses. Thus, the easier source of information on the 1933 *Census of American Business* and the 1935 *Census of Business* is found in the Appendixes of the 1940 *Census of Business,* vol. 2, *Wholesale Trade, 1939* (Washington, D.C.: U.S. Department of Commerce, Bureau of the Census, 1942), pp. 1037–49.

Table 1. Wholesale Trade Establishments, by Types

Year	Total Establishments	Merchant Wholesalers	Manufacturer's Sales Branches	Manufacturer's Sales Offices Without Stocks	Petroleum Bulk Stations and Terminals	Agents & Brokers	Assemblers, Primarily of Agricultural Products
			NUMBER OF ESTABLISHMENTS				
1929	168,262	79,178	16,863	–	19,587	18,467	34,143
1933	163,708	82,865	12,444	4,429	26,176	13,818	23,962
1935	176,756	88,931	11,541	4,065	27,333	18,147	26,515
1939	199,726	100,961	12,844	5,082	30,825	21,083	28,931
1948	243,366	146,518	15,716	8,052	29,451	24,361	19,268
1954	252,318	165,153	14,759	7,831	29,189	22,131	13,255
1958	287,043	190,492	25,240		30,520	26,666	14,125
1963	308,177	208,997	28,884		30,873	25,313	14,110
			PERCENTAGE OF ALL WHOLESALE ESTABLISHMENTS				
1929		47.06	10.02	–	11.64	10.98	20.29
1933		50.61	7.60	2.71	15.99	8.44	14.64
1935		50.31	6.53	2.30	15.46	10.27	15.00
1939		50.55	6.43	2.54	15.43	10.56	14.49
1948		60.20	6.46	3.31	12.10	10.01	7.92
1954		65.45	5.85	3.10	11.57	8.77	5.25
1958		66.36	8.79		10.63	9.29	4.92
1963		67.82	9.37		10.02	8.21	4.58

salers Proper, but maintained its earlier complement. Manufacturer-Owned Outlets became Manufacturers' Sales Branches and lost its more complex structure in favor of a simple dichotomy between such branches "with stocks" and those "without stocks." Other classes remained stable save for the successful integration of the oddments that had previously made up the "all other" group.

A second emergency census, conducted in 1935, saw little change in the categories but gave some attention to nomenclature. The largest class changed again, from Wholesalers Proper to a name only an ardent taxonomist could stomach—to Full-service and Limited-function Wholesalers; and for the first time, but permanently, warehouses for chainstores were shifted into the province of retail trade.

In the 1940 census the data for 1939 fell into the five general classes that have become standard. In the case of manufacturers' branches a distinction was made as to operation with or without stocks, which has persisted. The only qualification of this tally of 1939 that need be made is that the grim and inadequately truncated class Service and Limited-function Wholesalers remained.[7]

THE SEMIDECENNIAL WHOLESALING CENSUSES. The act of 1929 calling for a census of distribution, later renamed census of business, provided for an integration of that inquiry with the decennial census. The tallies of 1933 and 1935 were exceptional and were not repeated in the 1940s. But because the desire for a more frequent appraisal of economic activity had been firmly planted during the New Deal era, starting in 1948, business censuses have become basically semidecennial. The first postwar census was carried out in 1948 and was issued as the Census of Business of 1948, completely separate from the population and agriculture censuses. Subsequent business censuses have been taken in 1954, 1958, 1963, and 1967. It appears to be policy to settle on this quinquennial or quadrennial timing for future censuses.

Since 1948 the breakdown of wholesaling has remained constant, allowing us to tell far more about the dynamics of the trade than had been possible during the period of trial during the previous two decades. In 1948 the largest component of the trade came to be called Merchant Wholesalers instead of the infelicitous Full-service and Limited-function Wholesalers. With the standardization of criteria that grew out of the standard industrial classification, first utilized in the 1948 Census of Business and first published in the *Standard Industrial Classification Manual of 1949*, comparisons became safe as well as simple. The comparison of the number of establishments, by type of operation, and the sales percentage of each type presented in Table 1 are consistent for the last four censuses. Prior to 1948 there is less direct comparability growing out of differences in the system of classification of types due to the former inclusion of activities no longer classed as wholesaling, such as chainstore warehousing. Nonetheless, the table allows us to make reasonable measures of changes that have taken place in wholesale trade over the period of 34 years for which we have information in detail.

[7] Ibid., p. 11.

Changes in Wholesale Trade: 1929–63

When the first count of wholesaling establishments was made for 1929, 170,000 places of business were so classified. Growth in total number of establishments was slow during the economic lethargy of the 1930s and the economic disruption of the Second World War, so that the 1948 total was only 27.3 per cent above that of 1929. The fifteen years after 1948 witnessed a quickening of the rate of growth from 1.43 per cent per year (1929–48) to 2.84 per cent per year (1948–63), so that there were 42.6 per cent more wholesaling establishments in the last census than there were in the first conducted after the war.

The quickening pace is not uniform in its impact. The firms engaged entirely in wholesaling and the holding of stocks, which they themselves own (Merchant Wholesalers), have proliferated far more rapidly than any other group. The 70,000 merchant wholesalers of 1929 were in 1963 approximately three times as numerous, whereas the Manufacturers' Sales Branches showed only a 75 per cent increase, and other types of trade a smaller rate of expansion or an outright decline in numbers, as was the case with Assemblers of Farm Products. Some of these changes result from variations in the criteria of inclusion over the years, but during the period of relatively consistent standards (1948–63) the basic trends show up as strongly as in the whole, and somewhat less consistent, period. It is clear that the dynamic components of the trade are Merchant Wholesalers, Manufacturers' Sales Branches, and Petroleum Bulk Plants; and only Merchant Wholesalers now make a larger percentage contribution then they did in 1929.

Changes in sales contributions tell a rather different story. In 1929 total wholesale sales amounted to some $67 billion, a figure that actually declined during the depression of the 1930s, not having recovered until after the outbreak of the Second World War. By the war's end the total was growing rapidly, reaching $100 billion in 1948 and almost double that figure in 1963. Some of the "growth" was, of course, accounted for by changes in the value of the dollar. Applying the notion of the wholesale price index, we find that wholesale trade has grown by 281 per cent rather than the simple dollar growth of 435 per cent over 1929. During the same period, Merchant Wholesaler sales no more than kept pace with the growth in all sales, maintaining a 432 per cent growth 1929–63. But Manufactures' Sales Branches outpaced all other forms of wholesalers, increasing their sales by 713 per cent between 1929 and 1963. Admittedly, all these figures are approximations due to changes in the content of classes; but there can be no doubt about the magnitude of sales growth even if the figures lack absolute precision. Petroleum sales, as any American would predict, showed the most substantial rate of growth, 922 per cent between the first and most recent censuses. Merchandise Agents and Brokers' sales grew at a rate only two-thirds that of the average, 273 per cent for the period 1929–63, and Farm Products Assemblers dropped farther behind the general growth rate, securing an expansion in sales of only 110 per cent in the 34 year period.

The import of these differences in rates of growth shows up directly in the percentage contributions of the various types of establishment to the total of wholesale sales. Merchant Wholesalers have kept a constant share of the trade which stands at just less than half for the whole period. Manufacturers' Sales Branches have waxed fat over the period, raising their share by nearly ten percentage points to just less than one-third the total sales. Petroleum wholesalers and farm assemblers have figuratively swapped places, with the oil dealers after 34 years now having sales more than twice as great as the farm assemblers. Brokers and agents have gradually dropped in contribution so that today their sales are only two-thirds as great in percentage as they were in 1929.

This tells us, along with the information on numbers of establishments, that we must look to the burgeoning Merchant Wholesalers and the Manufacturers' Sales Branches for an understanding of the present dynamics of wholesaling. Petroleum selling grows, but the figures derive in considerable measure from the relative youth and underdeveloped state of the industry in 1929. Changes do occur in the two other types of wholesaling, Merchandise Agents and Brokers and Assemblers of Farm Products, but their progress is slow and somewhat wasting. Historically, agents and assemblers are more significant than topically would be the case.

Geography and Wholesaling

Up to this point, our view of wholesaling has remained essentially that limned by the Bureau of the Census. There are reasons for accepting the census taker's conclusions, notably the wealth of information he provides us, but if we are to step over the pitfall that is fact for its own sake, we must take what the enumerator offers and turn it to our own purpose, and that purpose is not simply collection. The type-of-operation [8] interest in wholesale trade that characterizes the census publications in the field is directly an outgrowth of the viewpoint of businessmen and business administration instructors.

In seeking to undertake wholesale trade, two matters become central: (1) the organization under which that trade will be carried on, and (2) the particular line of trade. The census provides information on these two, the first through the breakdown by type of operation, which we noted above, and the second through the extensive tables dealing with kind of business. Some of this breakdown is of value; some is not.

We may feel that Revzan, despite the considerable contribution in his full study of wholesaling, still leaves us to shape one of the three legs that make for stability when he tells us, "however viewed, whole-

[8] For a clear presentation of the breakdown of the trade in the census see Appendix D of the *1948 Census of Business,* vol. 5, *Wholesale Trade, Area Statistics* (Washington, D.C.: U.S. Department of Commerce, Bureau of the Census, 1951), pp. 51.14 and 51.15, wherein the two-fold interest is shown through a distinction between Type-of-Operation Groups, e.g., Merchant Wholesalers, and Kind of Business Classifications, e.g., general line grocery wholesalers.

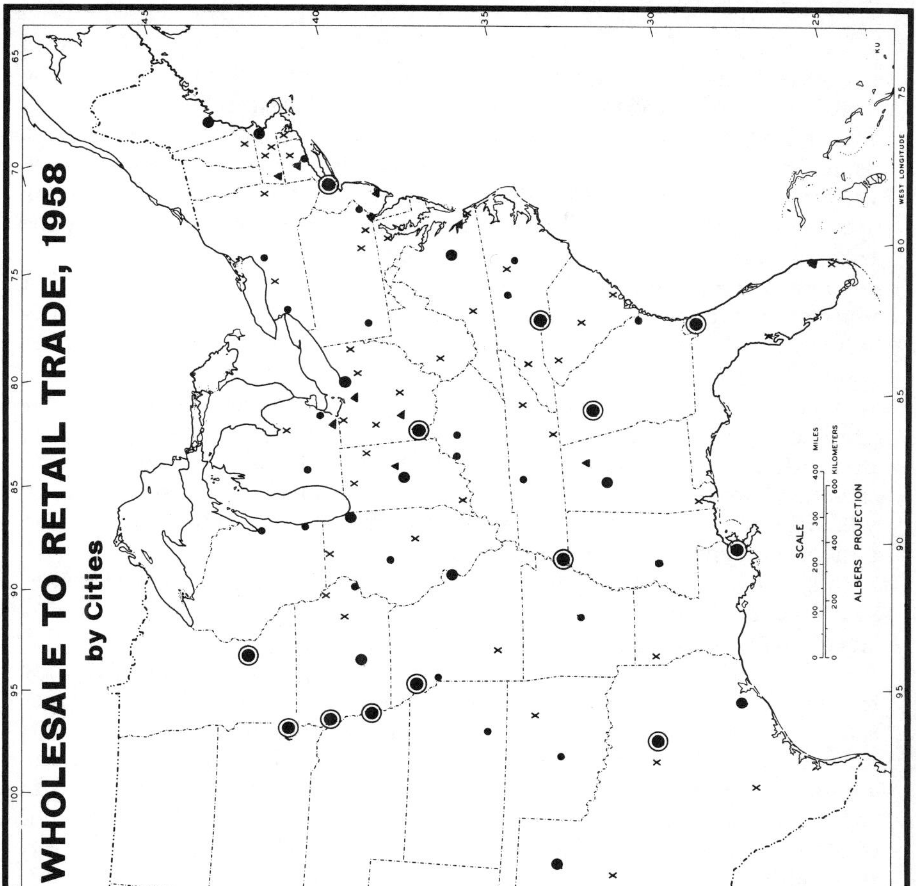
RATIO OF WHOLESALE TO RETAIL TRADE, 1958
by Cities
2.50 or more (13)
2.49-2.00 (14)
1.99-1.50 (28)
1.49-1.00 (49)
0.50 or less (13)
SCALE
MILES
KILOMETERS
ALBERS PROJECTION
WEST LONGITUDE

Figure 6

saling is a part of marketing which emphasizes matters of structural arrangement based upon degrees of specialization of both *types of business and kind of transaction.*" [9] He does not completely overlook the balance that is furnished by the leg that is an understanding of the location, but, as a professor of business administration, he tends to foreshorten it. In the section *Geographical Structure of Wholesaling* Revzan focuses his attention on the "geographical exchange of goods [which] will depend basically upon two sets of considerations. First, one area [whatever its economic–geographic size], which finds itself in a *deficit supply* position, must offer either a money price or its equivalent in goods and services, sufficient to generate a flow of the commodity needed from the surplus-producing areas." [10]

The geographical exchange takes place between "economic–geographic areas" that are political units of major size (states and census divisions). For the size and arrangement of these areas there is no *structural* explanation to go along with the channel concept through which Revzan furnishes the stuff from which the two legs—kind of transaction and type of business—are turned.[11]

The absence of any integrated and extensive consideration of *wholesale location* in previous study, other than in an organizational view of "location," provides a position for geographical inquiry. To know that New York, Illinois, California, Pennsylvania, and Ohio are the five most important states in wholesaling and account for almost half the trade does not provide an *understanding* of the location of wholesale activity particularly with respect to the processes shaping location patterns. Again, we must view the problem not merely as one of accounting for the location of wholesaling in national space, but as much more one of explaining where within the complex geographical area of the city that activity will be carried on. For the geographer, scale has a variable focus, and a clear image can be secured both from the local component and the over-all system. Some work has been done on the national distribution of wholesaling, but focused on urban, rather than the national systems.

The geographer examines an economic landscape in terms of its components and the activities they serve. He also seeks to understand the structure of the physical representation of activity that he sees. On the other hand, the scholar of business administration or the systems analyst may be satisfied when he has an understanding of the "organizational system," for each seeks only an operational understanding. But for the

[9] David A. Revzan, *Wholesaling in Marketing Organization* (New York: Wiley, 1961), p. 1. Italics supplied.

[10] Ibid., p. 75.

[11] In *Wholesaling,* one of two books on the subject (the other being the several editions of Beckman and Engle's *Wholesaling*), Revzan sets out his idea that the trade is a "channel" characterized by "linkages" that tie the producer to the consumer, and of "blockages" that impede free flow. In this view, the major concern with wholesaling is with its "organizational system" that explains "linkages" and "blockages." As an explanation of the institutional structure of wholesale trade the book is a distinguished contribution, though it leaves geographers to attempt to solve the location problem. See chap. 5, Ibid., pp. 107–42.

geographer whose interest is in the pattern and structure of economic activity as they occur in space, present-day operations are not enough. In the geographical study of wholesaling it is important to understand what the trade did when first established and how it has changed over the years, because the total spatial system may have been shaped by past conditions as much as by present objectives. If we wished only to know how wholesaling works, the present might serve; but since we wish to know *why* it works *where* it does, we need to know the past as well.

Not all types of wholesale trade pose distinct locational problems. Those men in the trade who deal in orders and bills of sale rather than goods, those who serve primarily as agents in buying and selling for their principals, and others classified as wholesalers who carry on essentially office functions, live by office locational standards, about which there is a fairly extensive literature. At the other end of the range are those wholesalers who stand somewhat in the class of pariahs in our society—junk dealers, renderers' agents, and petroleum bulk plant operators—and are therefore forced to isolate themselves from all save their peers. For these pariahs, solutions to the location problem are narrowly circumscribed by restrictive land use laws and ordinances and, for that reason, of less intellectual interest. A still different restriction on location decision comes in the case of agricultural and mineral collectors. To a varying degree their activities are located not by the general dynamics of wholesaling but rather by a special set of controls exercised by crop and mineral production. Interest of geographers attaches to the pattern produced in terms of agricultural geography, but that interest is discrete from the general problem of location of wholesale trade.

Under these circumstances four separate location problems emerge:

1. For the normal *merchant wholesaler* the location problem is complex, comprising a decision about the place of trade, the source of goods, and the sale of those commodities. The standard image of the wholesaler represents this merchant, and the traditional wholesaling district is his purlieu. At the same time, it is this merchant who has shown the sturdiest longevity and greatest ability to vary his location with changing economic and geographical conditions. In sum, the merchant wholesaler and his historical geography have the most to tell us about the conditions that have constantly transformed trade until the activity took the form it does today. For that reason, the greatest attention should be given to merchant wholesalers in any study of the geographical aspects of wholesale trade. No more than a variant form of the merchant wholesalers, when viewed in locational terms, is the *manufacturers' sales branch carrying stocks.* These establishments still must face true decisions about where to locate and to whom and in what manner to aim their selling. The source of supplies may be determinate in a way that is not the case for merchant wholesalers, but in other ways the establishment is sufficiently similar to his to be considered along with merchant wholesaling in a study concerned with economic location rather than business "organizational system."

   Those agents and buyers dealing in goods for which they take physical possession also live in the manner of merchant traders. It need not concern us greatly whether a commission agent owns the goods he stores in his warehouse so long as he handles those goods as a merchant wholesaler would. Obviously, commission merchants show a tendency to join together in the creation of an auction market, thus structuring their locational choice to a degree, but not so much as to throw them into a truly distinct locational class.

2. *Manufacturers' sales branches that do not maintain stocks* and many agents and brokers dealing in title rather than tangible goods form a second class of wholesaling. The sharp difference of scale from merchant wholesalers, manufacturers' branches with stocks, and commission merchants and agents brings distinct location dynamics. These men function in their trade in much the same way that salesmen on a circuit do, and sales branches might be thought the geographically fixed equivalent of the commercial traveler, or even as his pied-à-terre. In seeking access to the greatest number of customers "centrality" or "accessibility" are the paramount aspects of place, and the flow of men rather than commodities is their main concern. Since these same concerns motivate the central business district merchant and the financial district banker and broker, it is not surprising that the wholesaler needing nothing but an office usually ends up in the central business district or its retail replications in outlying areas.
3. The *pariah types* of wholesaling are controlled by negative forces to major degree. The oil tank farm is at the edge of the city because that was where society was reluctantly willing to suffer it at the time the bulk plant was built. Similar dynamics affect the location of scrap and waste dealers and others of the noisome trades. They cluster together in relation to their own, not any other wholesaler's, location controls.
4. Finally, *agricultural collectors* operate in a totally different context from that of other wholesalers, even the pariahs. The latter get as close to the city as they may, and the wholesalers in general normally choose urban locations, since their trade has at least a small relation to resident population totals. Agricultural collection shows no direct relation to population, even as it obviously shows a straightforward tie to the productivity of land. The broad nature of the tie in both cases may be that of a trade with its source of support. Beyond that tie, however, the similarity is slight. Additionally, many types of wholesaling serve the entire population, whereas agricultural collection is both limited in the geographical occurrence and demographically restricted. Thus, the locational characteristics of agricultural collection are distinct to that activity and susceptible to analysis in no wider frame.

This four-part breakdown of wholesaling, from a locational viewpoint, suggests that *in the study of merchant wholesalers is found the quintessence of the location theory for the trade.* Dynamics established in terms of merchant wholesaling cannot serve for all forms of the trade; but since the three other forms—office wholesalers, pariahs, and agricultural collectors—represent transitions between the archetypal form and retailing and office functions (for office wholesalers), primary production (for agricultural collectors), and a negative use of land (for the pariahs), knowing one of the poles should help to appraise the transitional form.

As geographers, we must seek to understand the location of wholesale trade in both its contexts—within the city and within the national economic space. That purpose is best accomplished by concentrating attention upon the physical pattern of wholesaling and from that pattern inductively establishing a structural theory capable of application over the timespan during which modern location decisions were made. In the following chapter, the growth of the national pattern is sketched to bring out the emergence of structure in terms of a nonuniform use of space, leading to the application of a new concept, that of *demand accumulation,* to trade location theory. Chapter 5 shifts focus to the urban area itself and applies to the analysis an additional dynamic notion, that of *decision abstraction,* seeking to integrate the location structure of wholesale trading within cities with that for the nation.

CHAPTER 4 *growth in the power of exchanging*

Adam Smith believed, "As it is the power of exchanging that gives occasion to the division of labour, so the extent of this division must always be limited by the extent of that power, or in other words, by the extent of the market. When the market is very small, no person can have any encouragement to dedicate himself entirely to one employment. . . ." [1]

While men were jacks-of-all-trades, they had neither the craving for, nor the substance with which, to engage in trade. Self-sufficiency implied the narrow horizon and the atomical region. Such minute dissection of the possible scope of trade could come either from the extreme fragmentation of political power or from the overly sparse population of regions. Feudalism may serve to identify the first and the pioneer economy the second. Each has produced a specific settlement pattern and a derivative theory of trading location. Feudalism provides the fundament for central-place theory, but the pioneer economy had not been taken fully into account until the present proposal of a mercantile theory of settlement.

Men could specialize in their labor; thus their wants would be served in part by the labor of others, and the mass of men so specialized in labor would provide an ever-growing market. Circularity would enter, and so long as the growth continued, both the parts and the ties would burgeon. In the absence of a reversal of expansive tendencies, increasing specialization of labor would lead to growing market areas and, in turn, enlarged market areas would foster labor specialization. Thus, the muscles of trade become the dynamics of change in the context of economic geography.

[1] Adam Smith, *The Wealth of Nations,* bk. I, chap. 3.

WHOLESALING AND SPECIALIZATION. Before looking at thos ics, however, it is necessary to ask two questions. First, does s tion affect the growth of trade as it does the labor of the in Phrased another way, this is the axiom reversed: Does phylogeny recapitulate ontogeny? Do we first have the all-encompassing activity and only later the narrowly defined trade? The answer to this question is not idle; on it rests the method of study and analysis to be used in the consideration of wholesaling.

Obviously wholesaling is a transitional activity, between production and consumption; but has it always been that, and is it possible that it may at one time have subsumed the far wider function of caring for all trading? If that broader undertaking were once typical of wholesale trade, then it seems entirely possible that wholesaling rather than retailing represents the fundamental form of trade envisaged by Adam Smith as implementing of the primordial division of labor.

It is the thesis of this study that such is precisely the case, that *wholesaling rather than retailing was the genesis of trade;* and that for the simple but powerful reason, to turn Adam Smith's equation around, that only through mass demand could labor specialization (division) be called forth. Retail needs are local and do not initiate extensive trade or labor division. Retailing serves merely to measure the possible extent of trade and to telegraph that quantity. Within a local market each producer could measure for himself and had no need to send any message. Thus, without wholesaling there would be no retailing divorced from actual production and, in a biological analogy, it is from the life history of trade, its ontogeny, that we find where in the developmental sequence of trade as a whole, retailing, the phylum, first appeared.

This sorting out of trading stages suggests the second question that must be asked: Can we generalize about the *aspects of growth* of individual trading centers, and their merchants, sufficiently to see the broad dynamics of trade? Here Adam Smith comes to our aid with his statement about trade as the sinews of specialization. All that we need add is that specialization has a temporal as well as vocational meaning. Men rather naturally become more specialized with the passage of time.

AN ARGUMENT IN FAVOR OF WHOLESALE TRADE AS THE PIONEER. Tradition has it that towns grew up as trading places in which men sold the produce of their labor, from fields or workshops, and away from which they returned home with their just wages. The customers were drawn from the town itself or the same closely associated countryside of agriculture and the rural industries. This is a description of what we might call a *localized economy* in which there is direct association of producers and purchasers. If we think of the mechanism of trade as the town market place, then it is probable that the sellers were mainly producers and only incidentally sellers, apportioning their time between these two functions, and the purchasers were also often sellers of different products. The meaning of this for our discussion of wholesale trade is that in such a primordial trading system there was little specialization of function and small value placed on unique skills. *The qualities of the trader were not rewarded as much as the qualities of the producer.* This is perhaps fitting because in a localized economy there was no great problem facing the trader. He lived in the market area he was serving, and he had constant contact with his customers, so it would take a man of small insight to fail as a trader.

In the next stage in the development of trade, specialization is the key to change. With the growth of special skills of production, the producer could earn his reward more directly through concentrating on the making of a single product. Also, he would earn his profit more surely by focusing his attention on *production* rather than on the trading function. With the increased scale of manufacture or crop raising that is implied by specialization and the earnings of unique skill, we know from Adam Smith's general axiom that the market must grow. Such growth could come from population expansion, but it would normally come faster and more certainly from the enlargement of the market area. In those terms, the localized economy would become only a part of the total economy and, probably, a decreasingly important segment. The skillful producer would be more interested in a link with an incipient wholesale merchant whose role as agent would more surely lead to an enlarged market-area and expanded sales than he would in a tie with an incipient retail specialist who could command no wider market than could the producer himself.

Once we have assumed the conditions that would destroy the dominance of the localized economy, we have created the conditions for wholesale trade. If we ask what effect the growth of the market area has on the nature of trade, the obvious conclusion is that it emphasizes the contribution of selling. The producer–seller, if he sought to continue both offices, would have to give an ever-increasing amount of his time to trading, were he to try to expand his market area to support the enlarged production of specialization. To sell more, he would produce less, and the system collapses. The only solution comes through the divorce of production from selling; in other words, in the introduction of the practice we have called *agency of trade.*

SPECIALIZATION THE AGENT'S ROLE. The agent of trade comes rather quickly to develop his own specialization which falls in the realm of the prescience that was previously so automatic. Knowing conditions at some distance from the source of production is also, in most cases, perceiving conditions at some time in the future. This was notably true in the past when much time and effort were required to carry goods even fairly modest distances. Then the trader had to solve a rather complex problem of cost of goods, cost of transport, prices of sale today, and probable changes in the price structure before the goods could be delivered at the distant market. If he solved the problem successfully, it would be a reward for his trading skills, not for his expertise in manufacture.

All this tells us that once the local economy of the periodic market is replaced by long-distance trading, the degree of trading skill needed is greatly expanded. It might at first be argued that that trading skill is really retail ability. Closer examination shows that manufacturing specialization will almost certainly produce merchants more akin to wholesalers than retailers. We must grant that in the past the distinction between wholesalers and retailers was less clear than it now is, but the contribution of these first merchants was that of specialized understanding of market conditions over a wide area, normally, for a rather narrow range of goods. There is nothing deterministic in that narrow range. Man's abilities are limited, and so it is reasonable to assume that, given great demands of anticipatory ability and knowledge of distant and changing markets, the trader will reduce the problem by simplifying the range of products in which he deals and for which he must make judgments. The well-recorded history of the peddler substantiates this view. In fact, as a result of dealing in extensive goods rather than specialized ones, the time and effort of the individual trader, which after all is a finite quantity, will tend to be used where the effort is least, that is, close to the area of production. In those terms, we are forced back on the local market system with little reason for the producer to utilize the offices of an agent.

WHOLESALING, THE PIONEER OF TRADE. The logical outgrowth of these assumptions is that the first trade *qua* trade was almost certainly more in the realm of wholesaling than in retailing, accepting that those words carry connotations of only limited application to the past. Thus, the development of specialization in manufacture or crop raising led fairly directly to the specialization in trading that was exemplified by the introduction of the system of trading agents. Once the trading agent was created, men emerged who gave their full attention to commerce, and whose time could be broken down into segments that might be used for trade in distinct places. That fact makes the notion of the central place a narrow concept on which to base explanations for the commercial origin and support of cities. Central place serves in the primordial state of localized economies, or in feudalism, when there probably was no occupational specialization in trade; but once trade as such is introduced, we need a surer structure than central-place theory, for example, gives us.

TIME OF DEMAND FULFILLMENT. Trade allowed the unusually adept man to utilize his skill, even if that competence lay outside agriculture or the most simple domestic production that had formed the support of localized economies. More important for us, trade gave the producer and merchant the chance to control to some degree the time of demand fulfillment. So long as men's needs were met by their own production, the perception of want tended to be followed immediately by its satisfaction. With the introduction of trading, the automatic response that was no longer possible on an individual level was maintained in the aggregate. The merchant became the mechanism that preserved that adjusting relationship of want and satisfaction by (1) moving goods from the place where they might be produced and in plentiful supply to the place where they were scarce (which is place utility), and (2) by anticipating the demand for goods either before it was expressed, allowing the goods to be available on demand, or judging the frequency and volume of demand sufficiently accurately to provide consistent employment for workers at their call (which we might term the utility of anticipation).

The Contribution of Demand Anticipation. To accomplish such adjustment required that the merchant be able to manipulate the market in place and in time. Merchants have always gained their profit, at least in part, from furnishing place utility. To understand the growth of wholesaling, we must add the notion that merchants have also profited from the contribution of *anticipation.* In the early stages of trade, anticipation takes place in terms of satisfaction of want periodically and at specific times. Such a parceling out of a merchant's time and capital allows him to collect several, probably individually submarginal demands into a single body of support that is profitable.

THE GEOGRAPHICAL SCALE A FUNCTION OF SPECIALIZATION: THE WHOLESALING AXIOM. Given the two functions of wholesaling, (1) moving goods from areas of plenty to areas of want and (2) anticipating times of possible oversupply and undersupply, several aspects of specialization are presented. The supply of goods may be specialized by type, time, or area of distribution. Given a small demand in a specific locale for a particular type of good, trade will be either highly infrequent or else characterized by "commodity combining." In the second case the merchant will make more frequent visits to the locale than his trade in a single item would warrant through the simple expedient of seeking to sell there a variety of goods, the total demand for which will make the step worthwhile. *If the merchant wished to become more specialized in his lines of trade, he would have to expand his trading area.* Or the reverse of this, *if he wished to narrow his trading area, he would have to broaden his lines of trade.* Only an expanding market would change the application of these rules. Given more customers in the same area, coming shall we say with settlement maturity, then specialization of line or specialization of trading area could be carried out.

MARKET SPECIALIZATION AND LOCATIONAL EQUILIBRIUM. This concept of specialization in the trading area for wholesale establishments is related to Lösch's idea of location equilibrium, which argues that there is relentless economic pressure to reduce the size of areas tributary to a particular center.[2] That pressure is an outgrowth of increases in total trade that open avenues for competitive trading centers to be set up nearer the customer. But, as we have already noted, *wholesale trade does not appear to be governed by competition among trading centers as much as by competition among wholesale establishments,* possibly all in the same trading center. The integration of that idea into a structural concept for the location of wholesaling activity requires a specific mean-

[2] The concept of locational equilibrium is briefly summarized in Chapter 2.

ing of the word specialization. Given a small market population, trade must be accumulated and expressed at infrequent intervals. This vitiates the substance of Lösch's idea, if not its precise statement. Smaller tributary areas, in his thinking, provided *better* service to the customer. In wholesale trade, more frequent *but less specialized* service may, in fact, be poorer service. *Thus, it is the quality of service not its proximity that is the key.*

However, we find that the dynamic nature of the tie of wholesaler and retailer conforms to the following restriction of Lösch's statement: Tributary areas must be as small as is consonant with the highest quality of service. *As the tributary area population increases, the wholesaler must increase his product specialization, and thereby improve his service, in order to retain the same area of trade.* If he does not increase his product specialization, then his tributary area will be forced to contract through the loss of integral parts to competitors who can prosper on no more than a part of an expanded market population. In these reduced tributary areas the frequency of trade would be increased at the expense of product specialization, which is a form of higher service.

These ideas suggest a structural generalization applicable to trade at various places and throughout modern times.

1. Wholesaling is a connective function under which the limits of location are the two poles represented by the source of goods and the point of their ultimate consumption.
2. Thus, the location of an individual wholesaling establishment will be determined by the interplay of time utility and place utility in which the customer desires immediate local access to goods and the wholesaler strives for the maintenance of tributary areas sufficiently large to allow him to establish reasonable forecasts of the timing and type of demand for goods.
3. Since neither the physical viewing of goods for purchase nor their transport by the customer is essential, the quality of wholesale service is normally more important than proximity of the merchant to his customer. In this regard, wholesaling differs strongly from retailing and is not governed by rules of gravitation.
4. The quality of wholesale service is comprised of time and product choice wherein the customer seeks the precise goods of his choice at the moment of his choosing, and the wholesaler the restriction of one of these types of utility. Thus, the wholesaler seeks either to provide immediate delivery of an unspecialized array of goods or the forward delivery of specialized goods.
5. As the total market support increases, product specialization and time continuity of trading may be increased, with the maximization of *service* determining which will take precedence.
6. The dynamic nature of the concept of service and the change in the total market support account for the existence of distinctive solutions to the problem of maximizing service and resulting contrasts in wholesale trade establishment locations.

It follows that the fundamental dynamic in wholesale location is the degree of time and place utility that may be furnished. Practicable limits on either of these utilities grow out of the interaction of specialization and tributary area size, recalling Adam Smith's axiom that larger tributary support permits enhanced specialization. All that has been added is the notion that specialization must be read as both a product and a time function.

THRESHOLD AND RANGE OF A GOOD IN THE CONTEXT OF WHOLESALING. Time specialization in trade made it possible for the merchant to secure the geographical separation of trading by calendar periods. This meant that one merchant could set out to serve a number of demands, which individually might be submarginal, because by collection they would provide an adequate support. The *trading circuit* cannot be appraised by using the notion of the threshold without some important modifications. This should not surprise us, for the idea of threshold was derived by Christaller, not from the most basic form of trade, but rather from a later and derivative form, which is retailing. In wholesaling, the fundamental form, the threshold has two expressions. Thresholds are, as Christaller said, a minimum amount of support that will bring an activity into being;[3] and, as he noted, the value of the threshold changes through time. In addition, the value of the threshold also depends upon continuous or periodic demand fulfillment; thus the second expression—the threshold is a minimum amount of trade, *either unitary or dispersed in expression,* necessary for the support of a particular merchant.

If the threshold concept needs to be refined, so does its obverse, the range of a good. Christaller defined the ultimate range of a good as the "upper limit" of its range. In this way, no one center could hope to collect and hold an amount of trade greater than the upper limit because, if it did so, it would introduce a condition which Lösch subsequently termed "conditional disequilibrium." A new place of trade would be established, and competition would reduce the trade to a figure no greater than that envisaged as the upper limit of the range of good. With a condition of peripatetic trading along a circuit, no one place by itself rises to the lower limit of the range (the threshold), but also no one place approaches the upper limit that we call the range. Thus, Lösch's idea of abnormal profits cannot come in directly, and *competition is introduced not geographically but by service and price warfare expressed from the same center the initial trader used,* for it is there that collection of submarginal units of trade is, in the geographical sense, accomplished.

THE TRADE CIRCUIT IS ADDED. It is clear that, so long as *trade circuits* exist, they tend to confer upon the initial trading place a power that limits the application of the notion of the range of a good. As Britain's ties spread, the London merchant effectively pushed his own trade far towards the limits of the empire. When competition arose, the second establishment was normally sited in London because the movement of a competitor into the trade may be made most easily and prof-

[3] This is Christaller's "lower limit" of the range of good, which has come to be called by American students "the threshold."

itably from the same center. Up to a point, this condition prevails. That point is the shift of trading from a periodic circuit to a constant tie of customer and wholesale merchant.

The circuit is never totally abandoned, but is modified in relation to an enlarging market. As that support grows, the two restrictions on the development of trade—the need to combine many types of trade to secure a marginal profit for the operator and the need to combine many possible markets, through periodic trading, to reach a threshold value—change. With increasing trade, product specialization becomes more readily obtainable, and, in fact, more desirable. As Adam Smith saw it:

> The dealer who can employ his whole stock in one single branch of business has an advantage of the same kind with the workman who can employ his whole labour in a single operation. As the latter acquires a dexterity which enables him, with the same two hands, to perform a much greater quantity of work; so the former acquires so easy and ready a method of transacting his business, of buying and disposing of his goods, that with the same capital, he can transact a much greater quantity of business. As the one can commonly afford his work a good deal cheaper, so the other can commonly afford his goods somewhat cheaper than if his stock and attention were both employed about a greater variety of objects.[4]

But obviously the development of specialization is an act of maturity rather than incipiency; the trader must be able to foresee a market of sufficient size to allow him to abandon the increments of support that come from diverse operations. With two variables at work—product specialization and time specialization—the problem of deciding how to conduct the business is not simple.

If the merchant decides to enhance his specialization by restricting his trading to a small group of products, he must, in a market of stable population and expenditure, expand his goegraphical horizon. Thus, *product specialization leads to geographical generalization.* But if he wishes to restrict his realm of trade, then equally he must broaden his array of goods. Thus, *areal specialization leads to product generalization.*

Adam Smith did not tell of this relationship, probably because like other Europeans he had little experience with the *pioneer economy.* His view of the trade was derived instead from evidence collected in a trading economy of considerable age and complexity. However, in 1776 when he was writing, the American colonies were still on the frontier of trade. Economic conflicts between metropolis and colony were central to the struggle just begun, and it should not be forgotten that the Americans were arguing for freedom of areal specialization, in contrast to the product specialization which was the more common in British wholesaling of the time. Adam Smith's maxim about the virtues of product specialization was, as an anthropologist would note, culturally relative.

What all this tells us is that the circuit continued, but its geographical scale varied with the degree of product specialization attempted.

[4] Smith, *Wealth of Nations,* bk. IV, chap. 5.

The traveler in lace ranged over major parts of the United States, while his brother in groceries fished intensively in little more than the backwaters of a county or two.[5] But the salesmen continued to ride circuits, giving to wholesaling a fundamentally different dynamic from the forces shaping retail patterns.

TIMELESS GENERALIZATIONS AND THE ECONOMIC STAGE. The structural concept organizing this brief historical survey of the growth of wholesaling is exactly the one alluded to before: that *wholesale trade represents a formalized and structured business relationship between a limited number of sellers and buyers,* and, thus, is strongly influenced by past experience crystallized into traditional practices. The ideas of a threshold and range of a good work differently in wholesale trade from the way they work in retail trade. The purchaser from a wholesale merchant is not necessarily either a buyer (in the retail sense of someone undertaking to look over goods and provide his own place utility) or a free agent (in the sense of being characterized by shopping for the goods themselves). The purchaser expects that the wholesale merchant will convey place utility by assembling goods from distant producers and transporting them to the site of the retailer's business. Because such movement normally implies a considerable amount of transport, the retailer does not normally set out to view the goods in the warehouse. The local shopkeeper expects the wholesaler to come to him with samples and to maintain some degree of uniformity in the product furnished in wholesale transactions. The merchant's creation of such a circuit of trading must, however, be based on logical expectations of profit. In that way there is an upper limit to the radial extent of a wholesale city's trading area, for eventually adding another peripheral ring of trade territory will add no more profit. So long as no countervailing trading place exists, the limitation of the range of wholesale goods is far more broadly drawn than is the case for retail merchants. The use of formalized selling links allows the wholesale trader consciously to determine his outer trade boundary. It is the marginal value of additional sales obtained at an excessive cost that finally draws the boundary rather than any absolute distance; and it is the wholesaler, perhaps more than his possible customer, who settles on the final boundary.

That marginality of one more sale gained at the extreme edge of a trading area is, of course, clearly tied to transportation and communications conditions. It is also tied to the stage of economic development. In a pioneer economy there may be no alternative to excessively expensive selling based upon warehouses in a distant center. The only other way would be to conduct wholesaling on an impossibly small scale at a closer point. In retail trade that "other way" is rational and normal

[5] Marshall Field and Company used two types of salesmen in their years as Chicago's largest dry goods wholesalers: "general salesmen" who sought sales for a broad line of goods in a restricted region, and men seeking to sell a specialty over a wide geographical area who were called "line-men." Lloyd Wendt and Herman Kogan, *Give the Lady What She Wants* (Chicago: Rand McNally, 1952), pp. 192–93.

because the customer views the goods and himself confers the final increment of place utility that is given to commodities, taking the goods from the local store to his isolated cabin. Individual selection and "do-it-yourself" place utility could be secured in no other way. In wholesaling, the traditionalization and formalization of trade make such geographical dismemberment of submarginal trading areas unnecessary. Even in times before the governmental protection of trade and traders and before the creation of large national economies, wholesale trade still could gain advantages of concentration through the periodic trading by wholesale merchants at fairs arranged in different places and at successive times throughout the year.

THE EFFECT OF VARIABLE PRICE ON THE LOCATION OF WHOLESALING. Another contrast between wholesale and retail trade that influences the very meaning of marginality is variable pricing. The retail merchant normally offers goods for sale at a fixed price. In the past that price customarily contained a masked charge for delivery, so at first it might be argued that the retailer also conferred the ultimate place utility. Closer examination shows that in fact the delivery charge had a test of normalcy—within the city, or some designated delivery area—beyond which an additional charge was levied. With the enhancement of the shopper's ability to provide his own final place utility, which came when purchasers usually shopped by car, the retail merchant could lower the price of his goods, and treat close and distant customers equally by abolishing delivery. The supermarket introduced this pricing policy in the 1930s and the discount house expanded it to many other lines of goods in the 1950s.

In wholesale trade, differences in the cost of carrying the goods to the customer [in providing the final place utility] were always subject to inclusion in the final price. Thus, the city retailer paid less for his goods deposited on his doorstep than did the general-store owner in the country. In this way, the wholesaler could then disregard the cost of delivery unless he wished to use a delivery allowance as a means of expanding his market or maintaining it against competition from wholesalers located closer to the customer. The price, varied by transportation charges, added to the base figure, allowed a diversity of locales to be served with some equity.

Another aspect of this system of variable pricing showed up in the encouragement of quantity selling in wholesale trade. To lead customers to buy in large units, quantity discounts were normal. This practice was one in stimulation of trade, it also was one rooted in the tradition of periodic selling that lay behind wholesaling. For the distant customer, his *conscious accumulation of demand* allowed him a price discount that could, in turn, be used to make up for higher final delivery charges. In this way wholesaling had a powerful economic weapon that could be used to defeat the exactions of the range of a good, or even in the time sense, the threshold. Policy made customary a variable price predicated upon delivery cost *and* size of unit purchased.

ARBITRAGE AND WHOLESALE LOCATION. To a very strong degree the conduct of wholesaling is rewarded by the profit of arbitrage, wherein value of a commodity at one place at a particular time will be greater than its value at another place. For the wholesaler to enjoy that profit, he must be able to conduct a business operating over a considerable geographical area. In the economic sense this means he confers place utility on the goods he moves outward toward the place where the price is normally higher. In addition, the wholesaler confers *time utility* in the sense that he holds goods in stock and furnishes them to the retailer both more quickly and more painlessly than that local merchant could get them for himself.

DEMAND ACCUMULATION STANDS AS A PIER. The time utility of wholesaling is not comprehended by Christaller's scheme, only its place utility contribution is integrated into central-place theory. Thus, we must add to the two piers that support central-place analysis—the threshold and the range of a good—a third pier which is the concept of *accumulation.* Because retail trade has what we may think of as a direct and personal tie with its customers, so that the multitude of individual decisions to seek goods or services produces an averaging of demand over a period of time, accumulation of demand by the individual is not very important. There are variations in trade during the week and among seasons, but the great number of customers of diverse interests gives an averaging effect. And for the individual, his demand for goods and services tends always to be relatively immediate and capable of satisfaction on recognition. Bear in mind that the central place–tributary area tie is one wherein each customer serves as his own agent for purchase, wherein he goes to the place of trade, and wherein it is rare for the customer to give more than a day to the journey-to-shop and return. But in wholesale trade the customer is not his own agent, does not normally physically go to the shop of the wholesaler, and gives no time to the purchase beyond the time taken up with the precise transaction.

There are exceptions to this, as we shall see, but as a generalization wholesaling transactions differ in these ways from those of retail trade. But to have another serve *locum tenens* in the execution of the transaction and the furnishing of place utility, it is necessary for that agent to have an economically significant transaction to carry out. At that point the often small and economically irrational demands of retail trade are ruled out, and demand must be expressed in large and less unpredictable units. The wholesaler's customer must accumulate his demand until it reaches a sufficient size to justify the transaction economically. The situation is rather like that that we find with electricity: a certain voltage is necessary before the power will arc across a gap, once the gap has been bridged, a new charge must be built up to repeat the process. Retail trade is a continuous flow along a line with little resistance, because of a short distance separating the customer from the merchant and an often noneconomic justification for movement to conduct business. Wholesale trade is a periodic undertaking wherein there is the accumulation of demand until it can produce a force sufficient to cross a considerable distance and in a manner such that the transaction must meet economic criteria before it will be carried out.

No one would doubt that place utility has great interest for geographers and their analysis of economic distributions. Time utility is not so obviously our concern. And yet the two concepts are closely related, and place utility is seldom totally divorced from the idea of being at a particular place at a specific time. The idea of arbitrage, for example, is not that one place always has a higher value for a particular commodity than another place, but more commonly that one place *at a given time has a higher value.*

The interrelation of place with time utility furnishes us an organizing concept to use in the analysis of economic distributions that depend upon *periodic* or *noncontinuous flows.* We know that retail trade can be thought to be controlled by the range of a good, with a certain minimum range needed for the trade to come into existence (the threshold), and the capacity to collect only a finite amount of trade at one single center (the range). But in periodic trade, such as wholesaling, it is easily demonstrable that the distances over which trade is carried on vastly exceeds those characteristic of retail trade. In many ways it almost seems that, although there may be many thresholds, there are no ultimate ranges. And still there are. What is their nature?

DEMAND ACCUMULATION AND CENTRAL-PLACE THEORY. The notion of *demand accumulation* provides a convenient measure for determining the specific weakness of Christaller's central-place theory in dealing with the pattern of wholesale trade. To show that weakness, we must examine the basic dynamics of establishment location choices as envisioned by Christaller.

In the framework of central-place theory, patterns of trade connections between potential customers and sellers are determined by the idea that it takes a minimum amount of collected trade to support any establishment. That minimum amount of trade is given a geographical expression through the use of the lower range of a good, which in the United States has come to be called the threshold population. The thinking is that a certain population in a localized area will permit the establishment of the first shop in a particular trade. An increase in that local population may subsequently permit the establishment of a second, and later of a third, and so forth. Christaller makes no comment on the period of trade, and therein lies the fatal flaw in the use of central-place theory to account for wholesale trade. We might point out that if we envisage a form of trade under which an itinerant peddler collects demands from various areas through periodic trading, the notion of the threshold is meaningless unless we add to it some idea of the timing of trade—whether it is infrequently periodic, rapidly repeating, or continuous.

The other support of central-place theory is the upper range of a good, which implies that for any single trading center there is an outer range of its tributary area, based on certain assumptions about the effort-cost to the customer in going to that center to secure the good. If, however, we make an assumption that trade is taken to the site of the customer, as in peddling or wholesaling, then the meaning of upper range changes. And even if we assume the continued gravitation of customers to the central place to shop, we cannot totally disregard the impact of time. We must appreciate that some shoppers "accumulate" their demand over a long time period and then use that heightened need to justify an unusually long journey-to-shop. Such accumulation makes the upper range a somewhat shaky concept.

Restated to comprehend three elements of analysis [lower range of a good (threshold), upper range of a good, and demand accumulation], this becomes trade-center theory:

1. A certain minimum population (as an expression of potential trade) is necessary to support any trading establishment. The size of the minimum needed is a function of the degree of demand accumulation practiced.
2. With great accumulation of demand, the least tributary area population will be required, whereas with the continuous or nearly continuous carrying on of trade, the maximum threshold population will be required.
3. In trade gravitation there is normally an outer limit to the range. But that outer range will be shortest with continuous trading and will be successively increased with reductions in the frequency of trading, which leads to accumulation of demand.
4. Phrased more succinctly, we may say that demand accumulation leads to a build-up in the desire for trade, which is expressed as the *impulse to trade*. For a given trading situation, the impulse to trade is least in continuous trading and greatest in most infrequent trading.
5. In all cases it is possible to envision the conduct of trade through the agency of an intermediary. Infrequent trading carried on over a great distance is most amenable to the intervention of such a trading agent, which in turn encourages the great radial increase of the upper range of a particular good.

By making these modifications of central-place theory, we create trade-center theory, which provides us a tool of great use in explaining the realities of wholesale trade.

THE FORCE BEHIND WHOLESALING: THE IMPULSE FOR TRADE. In this formulation there is, however, a missing element that is the impulse for trade.[6] In retailing that force is diffused among a multitude of customers and important only when expressed en masse. In wholesaling, individual impulses may be quite adequate to produce linkages over a very considerable distance. Therein is found the key to the explanation of wholesale establishment distributions. With a strong impulse, trade will be carried on (1) over a great distance, or (2) at a very frequent interval. With a weak impulse trade will either (1) be characterized by short linkages, or (2) be notably infrequent in execution. The value of the im-

[6] Impulse is defined by Merriam-Webster as "The product of the average force by the time during which it acts."

pulse will depend upon the absolute market to be found at one place and the duration of its accumulation. Obviously a large impulse could stem either from a mass demand and a short interval of accumulation or a smaller demand and a longer collection. But the geographical import of an impulse will come from its own momentum, no matter which means caused it to be. A given impulse will produce a link of a given value.

The foregoing is true because place utility is paired with time utility. Only when there has been an interval in trade [7] will place utility become apparent. It is granted that increasing experience and analytical understanding of trade will tend to make inventory holding more regular; but goods will remain constantly available only if there is a demonstrated utility in such continuity. If demands may be accumulated, economies of scale are introduced. In that way time utility has an essentially negative operation: the more infrequent the fulfillment of the demand, the more efficiently it may be satisfied. Thus, the customer who bought three suits at a time every five years would conserve his effort and that of the shopkeeper even if he might not conform to style, which is economically irrational. And the same man could afford to travel a goodly distance once every decade to buy six suits. Thus, to understand the operation of periodic trade, it is essential to add the concept of impulse for trade to those of the threshold and the range of good.

How Wholesaling Began

Perhaps the central condition in the development of trade over the centuries has been the slow emergence of consistency and specialization. The fact that Chinese products were available to the patrician Roman implies that the trade was geographically extensive. Charlesworth glows when he says, "The agents of the Roman business man had reached Ireland and touched the margin of the Baltic Sea, knew the Scythians of the Tauric Chersonese, and had met the Chinese traders beyond the lonely Stone Tower of Tashkurgan, had bought and sold in the marts of India, and bartered goods with the Aethiopians." [8] Those agents were of necessity self-sufficient men who were capable of setting out from Rome expecting to buy one commodity, could trade in a distant market for another product for which the time and place utility on the return to Rome could be expected to exceed that of the initial product, and might on the way home avail themselves of still other unanticipated opportunities for profitable transactions. Under such conditions, the careful distinction of different kinds of traders is artificial and meaningless. Opportunism was the rule of trade, and it may be well to use the old term "merchant" to denominate these early traders. Quite literally they

[7] Interval is defined by Merriam-Webster as a "space of time between the recurrence of similar conditions."

[8] M. P. Charlesworth, *Trade-Routes and Commerce of the Roman Empire* (Cambridge: Cambridge University Press, 1926), p. 239.

did "negotiate" and "traffic" in goods, often not caring greatly about the nature of their commodity so long as it yielded a profit.

These merchants were the ancestors of all present-day traders, but the family characteristics show most strongly among contemporary wholesalers. The conferral of place and time utility was the Roman merchant's contribution, and for it he was paid. He was not a shopkeeper, with all the consistency of place and purpose that that term suggests.

The opportunism of the ancient trader was his living but also his stigma, for society was not unmixed in its appraisal of the worth of anticipation of future demands. It was agreed that the merchants should be recompensed for the carriage of goods from distant markets. The exactions of transportation were obvious and relatively easily measured. But the value of forethought was much less amenable to easy calculation. The objection lay in the matter of profit as such. Philosophically the concept of profit was at first questioned, for it was thought that no natural limits existed.

> A mere cursory sampling of the obviously prominent spokesman of Greek and Roman philosophy (Plato) produces a consensus of mistrust of the merchant. . . . Although he conceded the useful and laudatory function of the merchant as a distributor of goods, nevertheless, he was pessimistic about the possibility of many men maintaining a moderate course in trade. Exposure to the opportunities of amassing wealth seduced most men to the temptations of extreme profits. Plato's solution was to curtail merchandising to the smallest possible scope, even to forbidding its practice to resident citizens of a city.[9]

This view, and its solution, was shared by Aristotle who "highly suspected the merchant's ability to withstand the temptations of greed. Since the merchant tended toward intemperance, which was destructive of virtue, he was disqualified from the privilege of citizenship in the best-governed state."[10] Only Cicero, viewing the situation at a later time, tempered the indictment, "granting a certain respectability to wholesale merchants" even though he "despised the small retailers because of their petty lies and frauds."[11] But for all the ancients trade was a thing to be watched in any well-ordered society. In a parallel with our time, as the Roman empire replaced the republic and society changed, merchants became respectable, and trade could flourish in the cities.[12] It could not have been otherwise, since rapid growth of the city of Rome made parochial economic production totally insufficient.

With the collapse of the Roman empire in the West, fragmentation did return and there was relatively little movement of merchants over any great distance. Before that ultimate decline of Rome the Fathers of the church had tended to codify the view that those who engaged

[9] John W. Baldwin, "The Medieval Theories of the Just Price." In *Transactions of the American Philosophical Society,* n.s., vol. 40, pt. 4 (1959), p. 12.

[10] Ibid., p. 13.

[11] Ibid., p. 13.

[12] Charlesworth, *Trade-Routes,* p. 236.

in trade were subject to unbearable temptations, and in the case of Ambrose of Milan [fourth century], "after praising the virtues and benefits of the labor of farmers, he violently denounced the merchant who 'farmed the farmer' in order to amass great wealth. 'Why do you change the industry of nature into fraud?' he demanded. 'Why do you diminish the abundance for the people? Why do you produce scarcities? . . . Your gain is the public's loss.' " [13] From this ecclesiastical thought grew the reluctance to allow virtue to merchants or to permit them to exercise their trade within the confines of towns. Along with this stricture went grave doubt about many kinds of profit, which was expressed by the term "*turpe lucrum* or shameful gain." [14] Although usury and similar exorbitant charges were clearly envisioned, there was in addition question about the whole system of trading itself. In the mind of the fathers of the church, the fundamental role of the merchant was questioned to the extent that merchants were severely restricted in their operations. Thus trade was small in scale because of the strictures of the church as well as the parochialism of feudal trading territories.

Only in the High Middle Ages, with the revision of church thought epitomized in the work of Thomas Aquinas, was the merchant partially restored to favor. It was accepted that moderate profit (*moderatum lucrum*) could be justified, but the problem was not totally solved in the sense that the measure of reasonableness still had to be worked out. Thomas nailed down the measure when after discussing the "just price" he noted that this was a value in which no more than a moderate profit was to be included. The moderate profit was measured essentially by using the "market price as the normative price with which fraudulent, usurious contracts of sale were to be compared." [15]

Under this view of proper trade, the current price in a normal market became equated with the "just price" for which the medieval church strove in its teachings. The tie caused the church, particularly where it held temporal as well as religious authority, to argue for the operation of markets under the shadow of the church and within its purview. The tempering hand of the public conscience was not without reason if we may judge from the case of St. Godric told by Pirenne. Starting life as a wonderer born in poverty in the city of Lincoln, Godric rose through beachcombing and peddling to the status of a merchant in the fairs of England. From the moneys he earned he was influential in setting up a company to load a ship to engage in trading along the coast of the North Sea. "The company prospered to the fullest. Its operations consisted in shipping abroad goods which were known to be scarce, and there picking up in return merchandise which it took care to dispose of in places where demand was the greatest and where might be realized, in consequence, the largest profits. At the end of several years this prudent custom of buying cheap and selling dear made of

[13] [Ambrose of Milan] *De officiis ministrorum,* bk. III, chap. 6, lines 16, 157, 158, trans. in Baldwin, *Just Price,* pp. 14–15.
[14] Ibid., p. 32.
[15] Ibid., p. 77.

Godric a very rich man." [16] It was just that custom which the medieval church decried and worked against. Godric's canonization came not from his prudence but rather from the renunciation of its fruits when he gave his wealth to the poor and became a hermit.

As well as the wish to have open trading in a normal market, the church, or the sovereign, during the Middle Ages might more strictly control the execution of trade by instituting regulated dealing wherein the quality of goods, their weight, and even their price, might be determined. In that case the current price, inferentially the "just price," was established by sanction rather than derived from bargaining. But to make such a system work it became imperative for trade to be carried on in designated places at carefully established times.

Most weekly markets enjoyed a limited geographical monopoly, and the seasonal fairs often were granted a dominance over a broad area. As privileges, these institutions could exist only through the awarding of a royal or church charter granting special rights. Of particular importance was the provision of what were perhaps first known as *pede-pulverosus* courts, literally, courts for men with dusty feet. In these subject and foreigner stood equal before the law, a situation untrue of courts in general, wherein foreigners normally had no status at all. Thus, a traveling merchant had no rights and no legal recourse in securing his property or just payments for goods sold if he traded outside the fairs. But in trading within a fair, the foreign merchant stood as the peer of the local tradesman and his equal in the special Piepowder Courts that convened for the duration and ordering of commerce during the fair.

To begin with the "Commercial Revolution," which some have perceived as taking place during the twelfth and thirteenth century,[17] was a time when trade had two strongly marked characteristics that have importance for the economic geographer. In the first place, for that time, as Abbott Usher noted a generation ago, "It is essential to remember, that medieval trade was after all comparatively small in volume; confined to a small number of commodities in any given region, and *periodic* rather than continuous. The great staple commodities found a market that was spatially extensive." [18] Second, that trade tended to be *extraurban,* either in the suburbs outside the city walls or in the open country, as in the case of the famous Stourbridge Fair in East Anglia. "The municipal constitutions tended, in practically all portions of medieval Europe, to raise obstacles to commercial development, but the traders found a larger freedom than would have been possible within the limits of the municipal constitutions in the fairs. . . ." [19]

[16] Henri Pirenne, *The Medieval City* (Garden City, N.Y. n.d., Doubleday & Co.), p. 82.

[17] Robert S. Lopez and Irving W. Raymond, "Medieval Trade in the Mediterranean World," in *Cambridge Economic History,* vol. 2, 155 ff. and 289 ff. (New York: Columbia University Press, 1955).

[18] Abbott Payson Usher, *An Introduction to the Industrial History of England* (Boston: Houghton Mifflin Co., 1920), p. 136. Italics supplied.

[19] Ibid., p. 136.

In the study of the geography of wholesaling the significance of these dual characteristics of medieval trade—periodicity and extraurbanism—is clear. The towns themselves tended to have markets, where the essentials of everyday life could be secured, "serving primarily as a bond between town and country." [20] The market was normally held one day in each week in the town to which the market privilege had been given. In unusual cases a "perpetual market" meeting every week day was held. Thus, the period of trade was short so that neither great place nor time utility was furnished by the market trader. In truth he was the forerunner of the retail merchant of staple commodities. But fairs were both less frequent and less widespread in occurrence. Large areas were given to fairs, and they were held normally for a relatively short period once during the year. "The fair, however, was not merely a basis for the retail distribution of the primary imports: the wholesale trade in the great staples of foreign commerce was likewise carried on in fairs." [21] The fair, to complete Usher's figure of speech on markets, was *the bond of one region with another region.*

The pent-up demand of the men living in one place for the goods to be secured from afar kept the merchants in business. "The merchant, following a tradition which went back to earliest times, was *an itinerant foreign trader;* he was not a local middleman who only supplied what was absolutely necessary in raw materials and goods that could not be manuafctured in the town, in order to keep the town's own economy going. [Because of the pent-up demand there was] the troubled complaint of the Lübeck shopkeepers to their council about 1400 that a Nuremberg merchant in Lübeck sold twenty times as much as any other, and as much in one day as a Lübecker did in a year!" [22]

ENGLISH WHOLESALING AT THE TIME OF AMERICAN COLONIZATION. The expansion of the role of the merchant, as opposed to that of the local producer selling his product to his everyday associates, is well brought out by the increasing demand for information about fairs distant from the wholesaling towns. In *The Traveller's and Chapman's Daily Instructor* for 1705 [23] the group of fairs shown below is listed, by place and specific time period, for England and Wales, Scotland, and Ireland. Although many of these fairs seem (from internal evidence) to have been quite parochial undertakings, goods from outside were sold and fleeces and other agricultural products were collected for transmission to others through the channels of trade. In this we can observe a clear transition from the local economy, represented by the market, to our more modern system of fixed, continuous trading in central places. The fair as it existed at the beginning of the eighteenth century, that is at the time of the English and French settlement of North America, did

[20] Ibid., p. 138.
[21] Ibid., p. 139.
[22] Fritz Rörig, *The Medieval Town* (Berkeley: Univ. of California Press, 1967), pp. 146–47. [Emphasis supplied]
[23] *The Traveller's and Chapman's Daily Instructor* (London: 1705, G. Sawbridge at the Three Flower-de-Luces in Little Britain).

Month	*England & Wales*	*Scotland*	*Ireland*
	FIXED FAIRS		
Jan.	15	10	2
Feb.	49	15	6
March	67	24	28
April	109	12	22
May	164	31	34
June	203	101	68
July	210	38	36
Aug.	233	61	70
Sept.	262	67	60
Oct.	172	101	70
Nov.	161	58	38
Dec.	47	26	13
TOTAL	1692	544	447
	MOVEABLE FAIRS		
	363	43	74

many things, but most obviously it was of great interest to "travellers and chapmen" coming from far beyond the normal range of a good. By coming those distances the wholesaler could shape a geography of trade more extended and complex than is sometimes realized.

Perhaps the best description of a fair that survives is that contained in Defoe's *A Tour Thro' the Whole ISLAND of GREAT BRITAIN,* first published in 1724, wherein he tells of the nature of Stourbridge Fair just before its decline in the eighteenth century.[24] The Fair was kept in "a large Cornfield" of about half a mile square located between the road from Cambridge to Newmarket in East Anglia and the River Cam. "If the Field be not cleared of the Corn [probably wheat] before a certain Day in August, the Fairkeepers may trample it under-foot, to build their Booths or Tents. On the other-hand, to balance that Severity, if the Fair-keepers have not cleared the Field by another certain Day in September, the Ploughmen may re-enter with Plough and Cart, and overthrow all into the Dirt. . . ."[25] In the fair "the Shops are placed in Rows like Streets, whereof one is called Cheapside; and here, as in several other Streets, are all Sorts of Traders, who sell by Retail, and come chiefly from London. Here may be seen Goldsmiths, Toymen, Brasiers, Turners, Milaners, Haberdashers, Hatters, Mercers, Drapers, Pewterers, China-warehouses, and, in a word, all Trades that can be found in London; with Coffeehouses, Taverns, and Eating-houses, in great Numbers; and

24 The quotations are taken from the fifth edition of *A Tour* published in four volumes in London in 1753. In Vol. I of that edition this description is contained. To still the cries of orthographic pedantry that affect many British scholars, it should be pointed out that "woolen" is Defoe's spelling.

25 Ibid., vol. I, p. 90.

all kept in Tents and Booths." [26] Intersecting this "Retail Street" and, thus, "parallel with the Road are like Rows of Booths, but somewhat larger, and more intermingled with Wholesale Dealers." The "Wholesale Street" in its turn is adjacent to the Duddery, a square bounded by very large establishments "peculiar to the Wholesale Dealers in the Woolen Manufacture." Defoe notes an interesting anticipation of the modern manufacturer's sales agency when he tells us,

> In this Duddery, as I have been informed, have been sold 100,000 Pounds-worth of Woolen Manufactures in less than a Week's time; besides the prodigious Trade carried on here by Wholesalesmen from London, and all Parts of England, who transact their Business wholly in their Pocket-books; and meeting their Chapmen from all Parts, make up their accounts, receive Money chiefly in Bills, and take Orders. These, they say, exceed by far the Sale of Goods actually brought to the Fair, and delivered in Kind; it being frequent for the London Wholesalemen to carry back Orders from the Dealers, for 10,000 Pounds-worth of Goods a Man, and some much more. This especially respects those People, who deal in heavy Goods, as Wholesale Grocers, Salters, Brasiers, Iron-merchants, Wine-merchants, and the like. . . .[27]

The trade that would today fall to the merchant wholesaler was also joined by what, in our terms, would be commodity collection trading. Defoe notes that as the retail and general wholesale function of Stourbridge Fair began to wane, its activity, and even its place, was taken by trading in wool staple and hops. "There is scarce any Price fixed for Hops in England, till they know how they fell at Sturbridge-fair." [28] It appears that at least the hops were brought to the Fair, but we may assume that Stourbridge was not simply a point on the line of supply connecting the growers of hops in Kent or Surrey and the vast number of tavern keepers, who then tended to brew for their own needs as they arose, but also what we would call an "auction market" in which prices were set.

Taking a measure of our understanding of the growth of wholesale trade as shown by Stourbridge Fair, we can say that the periodicity of trade continued, and specialization of function had already begun. In addition to the retailers, who came in part to serve the vast number of customers at the *Fair,* there were a number of men whom we would class today as merchant wholesalers with stocks of goods for immediate distribution. They afforded both place and time utility to their customers. There were as well the "manufacturer's sales agents" we have noted, who undertook to furnish place utility at some future time. And finally there were the agents, brokers, and commission merchants dealing in staples and establishing prices in a true market situation. Already there was a

[26] Ibid., vol. I, p. 90. Cheapside was at that time the main street of London's retail shopping area. The Toymen referred to were dealers in small metal "toys" rather than necessarily merchants of playthings.

[27] Ibid., p. 91.

[28] Ibid., p. 92. Defoe spelled it as it is pronounced, as Sturbridge.

slight segregation of function within the teeming mass of the Fair, either in the several streets and squares or in the several intervals of trade during the course of the Fair. But, transience was the key for all, and we may assume that only through collection over the period of a year, for expression during a single week in September, was the impulse to trade strong enough to impel goods and agents over the barriers of space, between London and other emporia and Cambridge. The rapid decline of Stourbridge Fair in the first part of the eighteenth century, which Defoe cites, may have resulted in part from the improvement in travel consequent upon the turnpiking of the Great North Road, which allowed the impulse to trade to reach a critical size more readily than had been true under medieval conditions of travel. As a generalization we may conclude by saying that *as travel and transportation are facilitated, the amount of impulse needed to produce trade is decreased and the probability of frequent or continuous trade is enhanced.*

The Wholesaling Brought to America

Any explanation of the geography of wholesaling that seeks general acceptance must be valid for a major economic system, not merely for a particular isolated area and time. Nonetheless, examples must be both specific with respect to time and localized with respect to occurrence. In a study for an American audience done by an American there is much value in shifting the locale of study to the United States and Canada. As we have already noted in the discussion of Christaller's theories, the pioneering economy is one that has interest for us even at the present time, and Europe experienced its frontier of trade so long ago that even epigraphy offers little upon which to base structural theory.

AMERICA THE PICKET-LINE OF TRADE. Viewed from the metropolitan country, colonies were investments in trade and, as such, an economic undertaking. The American colonies, both in the United States and in Canada, were planted as pickets of a mercantilist system to secure overseas resources and crops that were scarce or unobtainable in the home country. This mercantilist system epitomized the formal and traditionalized tie which we have already noted is so characteristic of wholesale trade. Once the overseas colonies were planted, the flow of their products and of the manufactures with which they were paid was legally controlled. Even in the case of products that were unobtainable from the home country, there was the provision that such goods must be imported from their foreign source into the metropolitan nation and then forwarded, usually in native bottoms, to the colonies. Even if "drawbacks" of duty paid on admission to the home country were allowed on goods subsequently forwarded to the colonies, the carrying trade and the agency of dispatch were sources of considerable mercantile earnings in the homeland.

For Americans, taking note of the clearly economic motivation of the early plantations is not idle. Our history has taught for generations that we are a nation of libertarians and protesters. Yet if we reflect on the matter, no one could seriously believe that any nation sets its dissenters up in freedom and relative plenty. Obviously the American colonies, in this case both British and French, were a creation of the first well-developed wholesaling structure.

THE MERCANTILIST COLONIES. Along with the development of commercial ties went the growth of the notion of mercantilism as national policy. As Adam Smith tells us, there are two ways of providing revenue or subsistence for the people and for the government, "one may be called the system of commerce, the other that of agriculture." [29]

Mercantilism calls for maintaining a favorable balance of trade not through free trading but rather through a market structured by regulations and mercantile monopolies. The end of feudalism at first seemed to encourage trade by creating large political units inside which commodity flows were freed from the strictures that had made the fair essential to medieval commerce. But the creation of unified countries also established nationalism and particularly restrictions in importation and exportation. For economic growth changes were needed.

> The discovery of America, however, certainly made a most essential one. By opening a new inexhaustible market to all the commodities of Europe, it gave occasion to new divisions of labour and improvement of art, which, in the narrow circle of ancient commerce, could never have taken place for want of a market to take off the greater part of their produce. . . . The commodities of Europe were almost all new to America, and many of those of America were new to Europe. A new set of exchanges, therefore, began to take place which had never been thought of before, and which should naturally have proved as advantageous to the new as it certainly did to the old continent.[30]

One of the critical facts of American historical geography is the simple fact that the system of mercantilism was well established before that of national emigration. Both Britain and France turned to New World trade before political and religious dissent encouraged men to leave their homes. Fish and furs had brought Europeans to North America in summer by the year 1500. Under those conditions the impulse to trade was sufficient to induce seasonal ties at a stage when the schisms had not yet led men into the wilderness for settlement. In Virginia and in eastern Canada "plantations" of men were made just after the turn of the seventeenth century, but, oddly, although these were largely economic in origin, they were weak in accomplishment. Perhaps only the emergence of a prime trade item—tobacco in Virginia or beaver pelts in the St. Lawrence valley—saved these economic creations.

The scarcity of such a staple in Acadia left the French colony there feeble and crotchety and completely unimportant to the world at large, despite its primacy in age. Apart from a most unfortunately small

[29] Adam Smith, *The Wealth of Nations,* Introduction to bk IV.
[30] Ibid., bk IV, ch. 1.

Huguenot group in the St. Lawrence valley, which was soon persecuted and destroyed by the fervor of the Catholic counter-reformation, French Canada lacked the sturdy base of merchants to support its growth. The result was one of virtually complete dependency on France. The Laurentian settlement was long unable to feed itself and was mainly an outpost from which Indian traders set off into the forest to deal with native trappers for beaver pelts, which were in turn sent back to Paris, and from which the revenue and support of the colony stemmed. In the Newfoundland fisheries the details differed, but the mechanics were the same. European outposts were set up around which collectors, fishermen in this case, gathered wild products which were then forwarded to Europe. The fishery, mainly because of the ease of transport in the industry, produced no more than periodic settlement until 1610, leaving the rocky shores uninhabited in winter and even less clearly the base for eventual settlement than were the fur stations up-river.

NEW ENGLAND MERCANTILE OUTPOST. New England was modeled from different clay. Although the land and trading rights, as in other colonies, had been granted to companies of merchants in the West of England and in London, the first serious settlers to touch the New England shore were willing immigrants. At Plymouth in 1620 and in Massachusetts Bay Colony from 1622 on, the privilege exercised was that of the trading company, but the work was that of men who brought their whole thoughts with them to America—of men who had left England for good. Their intellectual baggage included both experience and agreement with mercantilist notions. If commerce sustained the English prosperity, then it should serve equally the New England commonwealth. The staples that might have diverted attention away from trade were less engaging in New England than in Virginia, and the urban origin of so many of the dissenters, particularly those in Massachusetts Bay, encouraged a mercantile rather than an agricultural pursuit of revenue and support. The lack of any staple such as tobacco or furs in New England left the colonies without a clear future until it was discovered that trade need not necessarily focus only on an old and imperial state. Much as the Puritan conscience was at times troubled by the materialism of trade, the view that God's favor would shine on the elect came ultimately to make of the merchant's success proof of his election.

Boston ships began to scout out trade in the West Indies, in the Mediterranean, and in Africa, as well as in the now aging fishing colonies of Newfoundland. Innes states,

> trade began with the southern colonies and in their staples, and with the British West Indies, which started in the second half of the seventeenth century to produce sugar as a staple product. With the importation of slaves from Africa, an additional market was opened for dried cod, especially the poorer grades, and for agricultural products and lumber and, in the transport of these products, for ships. New England had developed an important trading nucleus with characteristics similar to those of England and with the same advantage of cheap, year-round ocean transport. In certain markets such

as the West Indies and indeed in Newfoundland she had the advantage of location over England.[31]

THE IMPULSE TO TRADE IN NEW ENGLAND. The notion of the impulse to trade makes the piracy of the Newfoundland trade by New England easily understandable. The relative proximity of Boston and Salem to the Avalon Peninsula was such that far less demand need be accumulated for trade with the Yankees than was needed for trans-Atlantic dealings. The genius of New England trade throughout most of the history was that of keeping costs low so that the marginal trade might be engaged in. Samuel Eliot Morison has said that, even in the early nineteenth century, the coastal towns of Massachusetts could out-trade the old mercantile nations by building ships cheaply and well and operating them without the inordinate land-based overhead that characterized British trading companies. In the seventeenth century the advantage of the spare, swift local merchant was just relatively greater because the impulse to trade was weaker and took more collection to force ties across the North Atlantic. But "the secret of maritime [and mercantile] success was that persistent enterprise which led her merchant-shipowners to 'trye all ports' and to risk all freights." [32]

Other ties also helped the Yankee merchants. The Puritan merchant may have stayed in London or he may have emigrated to Boston, but his experience was the same.

> Lacking the means by which to judge the financial capacity and business acumen of strangers, potential English creditors could rely only on personal ties to men in New England. Blood relationships between English suppliers and New England merchants were an exceptionally useful bond. . . . If not related to their English creditors, the first New England importers must themselves have had previous experience in business and have left behind friendships and reputations which they could draw upon in attempting to finance their new ventures.[33]

It was fortunate indeed in the shaping of the Yankee, with his drive and mercantile abilities, that Puritanism gained particular ground in the ranks of English merchants, as Quakerism did a generation later. Persecution thereby was the agent of commercial pioneering in the thirteen colonies, as it would have been in the St. Lawrence settlements save for the parallel export of the persecution itself. In the end, the Huguenots, who had made similar inroads in the French merchant class, were to be found only in cities in the thirteen colonies such as Boston, where Peter Faneuil and Paul Revere are still well remembered.

[31] Harold A. Innes, "Economic History of the Maritimes, Including Newfoundland and New England," in Innis, *Essays in Canadian Economic History,* ed. Mary Q. Innis (Toronto: University of Toronto Press, 1956), p. 35.

[32] Samuel Eliot Morison, *The Maritime History of Massachusetts, 1783–1860* (Boston: Houghton Mifflin Co., 1961), p. 18. For a clear exposition of Boston's trade role see also chap. 15.

[33] Bernard Bailyn, *The New England Merchants in the Seventeenth Century* (Gloucester, Mass.: Peter Smith, 1955), pp. 34–35.

Quite literally, trade became the *staple* of New England, and that trade was not mean for its time. "To be a merchant in Boston in 1660 meant to be engaged, wholesale and retail, in the exchange of a great variety of goods, to be ready to accept payment in all sorts of unexpected commodities and currencies, always to be seeking new markets in which to sell new kinds of goods and new kinds of goods to satisfy new markets. Versatility was one of the keys to success; to specialize was to decline." [34] And another face of versatility is diversity; so we find in colonial American trade the joining together of a number of demands to produce the marginal impulse to trade. Numerous descriptions of the aggressive Yankee captain trading to the Spanish Main as well as the British West Indies tell us that his activities ranged widely over the spectrum of human desire. By selling rum, slaves, ice, or passage to planters to spend the cool summer in Newport he managed to get by, sometimes in a rather handsome fashion. But to make journeys that others disdained or undertook less frequently the Yankee "tried all ports" and handled any commodity.

The Beginning of Domestic Commerce

Although the Yankee did some things in foreign parts that he did not do at home, trade did not fall under any such restraint. The same inclination that drove New England captains into poor anchorages in the West Indies, slave forts on the Guinea Coast, and, later, the trading ports of China, also led his landward brother into the interior. While the Laurentian French were creating a very specific trade with the Indians of the Great Lakes and Ohio valley, the Yankees were more modest in geographical sights but far more grand in mercantile ones. Sharing the views later to be expressed by Adam Smith, the New England colonists sought the mercantile road to support and prosperity as no other colony ever has and gave the United States the economic attitudes that shaped our history. Arthur Meier Schlesinger saw in these merchants the force that led first to the assertion of trading independence and second to the forceful securing of political freedom. At times the Montréal merchants toyed with such notions, particularly in the 1830s, but their lesser numbers and importance failed to carry the day.

What sort of men were the New England merchants and how did they operate? To understand their place we must recall that the original purpose of the British American colonies had been as a source of staples used in the support of the English people and their industry. The reflux of this easterly bearing tide of staples, which reached massive proportions before the Revolution, was an even more valuable flow of goods, notably dry goods but including many other manufactures, that came into American ports and moved on to the staple-producing countryside. Imbalance between the two flows was made up by shipments of specie to the creditor, which was largely Britain. To carry on the trade, then,

34 Ibid., p. 100.

it was necessary to keep up the flood of staple and specie, and to accomplish that end shipping and trading were needed. In the middle and southern colonies there were conditions that brought forth staples—flour, flaxseed, and meat from Jersey and Pennsylvania; tobacco from Virginia and Maryland; pitch, tar, and turpentine from North Carolina; rice and indigo from South Carolina; and timber from Georgia. But from New York northward there were no such bountiful products. The fish of New England could be duplicated by Britain's own West Country fishermen. The upshot was that the "enterprising merchants of New England developed a network of trade routes that covered well-nigh half the world. Possessing within themselves no staple with which to make returns for their vast consumption of English drygoods and other wares, the main resources of trade of these provinces were the fisheries, the molasses-rum trade, the marketing of slaves and the coastwise traffic." [35]

To a degree that few other areas have, New England had to trade to live. That fact traditionally is accepted and the trade assigned to distant areas, but reflection tells us that the "vast consumption of English drygoods and other wares": was not for the Yankees alone. As a community of merchants they did eat and drink and put on their own goods. Yet even more they carried those goods to the ultimate consumer in the other thirteen colonies, as well as in Nova Scotia, Newfoundland, and the West Indies. For the Yankee the staple by which he survived was the earnings of trade with the place of the activity being of relatively little concern.

The method of increasing earnings was not completely in the New England merchants' hands so long as he was the useful, but somewhat chained, intermediary between the tobacco planters of Virginia and the leaf merchants in London. If, however, the Yankee turned to activities where he could influence the scale of trade, say in the molasses-rum trade or in slaving, then he was free to encourage his own future. Even more, if he carried the dry goods and other wares into the countryside, stimulating a demand for the merchandise and controlling its supply and price, then even more he advanced his prosperity. Thus, the Yankee became both the ocean trader and the back-country peddler.

THE YANKEE PEDDLER. We have seen that the "initiative" of the Yankee captains lay in efficient operation, absence of preconception, and a small scale that allowed the impulse to trade to be expressed frequently and from relatively isolated harbors. On land, the impact of the Yankee was fully as fundamental in shaping the present economic geography, but less well documented. The picture of the Yankee sailor in Fu-chou has a romantic appeal that the dusty packman in early Ohio cannot call forth.

If specialization is the gospel of our time, its pronouncement was in the future for the earlier American. Thus, to try to distinguish "the wholesale trade of colonial America" is putting the ideas of a man in the head

[35] Arthur Meier Schlesinger, *Colonial Merchants and the American Revolution, 1763–1776* (New York: Barnes & Noble facsimile of 1917 ed., 1939), p. 24.

of a boy. There was wholesaling, as we know it, but for the merchant of 1800 there could be no thought of "wholesale only." He had two characteristics of operation that we must recognize. The merchant before and just after the Revolution was a capitalist in the precise sense that he had capital that he employed to maximize his earnings. If the largest return came from foreign adventures in trade, then he was an importer. But if shipping goods into the interior looked more remunerative, then he was a wholesaler,, and possibly an importer as well. The colonial merchant had little or no organization as we understand the term today. He had very few permanent employees and little regard for institutionalizing his trade. If the term wheeler and dealer did not carry such connotations of social irresponsibility and political immorality, we could use the term to good office as a description of the colonial merchant. He could turn and adapt as the occasion warranted, and above all he dealt in commodities.

THE WHOLESALER MOVES INLAND; THE IMPULSE TO TRADE, OUTWARD. Under these conditions trade was a discontinuous undertaking to begin with, save for a few staple products such as nails, salt, sugar, wine, rum, and other products that could stand travel and storage. The coastal merchant imported these goods and forwarded them to country storekeepers, who in the beginning traveled the short distance to Boston or Philadelphia to look over the goods he was buying from the importing merchant. And through that visit he identified himself and his standing. Margaret Martin furnishes us with a good description of just how such country merchants worked.[36] Farmers in the Connecticut Valley bought the relatively small number and amount of goods they needed in exchange for excess agricultural produce.

Wheat, corn, oats, and rye; butter, cheese, lard, tallow, flaxseed; bees-wax; lumber and lumber products such as shingles or staves, hoops and headings for making barrels and hogsheads; horses, cattle, and hogs; potatoes and vegetables; apples and apple cider—all were accepted in greater or lesser amounts by the country storekeeper, or else taken directly by the farmer to the merchants in the river towns. The farmer also made a few products directly for export such as pot and pearl ashes. As the country produce came in, in driblets for the most part, the merchant stored it, and, in some cases, transformed it. In particular cattle and hogs were slaughtered and pickled in brine to make the famous "salted provisions," great amounts of which found their way to the West Indies. Fish were pickled and flaxseed was cleaned or sometimes crushed to make linseed oil.[37]

Obviously the merchant could not hope for the neat and institutionalized type of trading we have today. To sell goods in the country he had to accept goods in return. Thus, not only did he have to forward manufactures but also he had to collect and forward country produce.

[36] Margaret E. Martin, *Merchants and Trade of the Connecticut River Valley, 1750–1820.* Smith College Studies in History, vol. 24 (Northampton, Mass., 1938–1939).

[37] Ibid., pp. 5–6.

In this way agricultural collection, which we have seen forms a part of wholesaling even today, is quite traditional. But even more, we find in this case the complementary aspect of the impulse to trade. The country merchant could enjoy a return on his capital (used to finance his general operation and to buy the sale goods he traded for country produce) through his ability to gather and store products *until he had a sufficient amount to justify a trading journey.* Also the country merchant did undertake certain very limited processing and repackaging operations that made *forward* shipment both easier and of more likely profit. That forward shipment was basically the result of the impulse to trade, but in this case "forward" in the opposite direction from the trade in manufactures. Again, volume and demand had to be accumulated before the trading act could take place.

The importance of wholesale trade seems generally to have been overlooked partly because the dividing lines then as now were vague, but also because we do not tend even today to think much about wholesaling as an activity. In Table 2, Margaret Martin gives us information about the main trading town of the Connecticut Valley.

Table 2. Hartford Merchants Classified by Type of Trade

Period	*Wholesale Only*	*Wholesale & Retail*	*Total*
Jan. 1765–Dec. 1765		12	12
Jan. 1780–Dec. 1780	2	51	53
Jan. 1795–Dec. 1795	19	68	87
Jan. 1810–Dec. 1810	32	50	82
Jan. 1819–Dec. 1819	38	34	72
Jan. 1829–Dec. 1829	36	33	69

Source: Martin, *Merchants and Trade,* p. 13. Data were taken from newspaper advertisement descriptions. All retail stores are listed under wholesale and retail since in some cases classification was impossible. Thus, the wholesale and retail category may somewhat exaggerate the wholesaling element.

What the tabular figures tell us is that throughout most of the late colonial and early national period wholesaling was a very important up-country activity. It is particularly interesting to note that the onset of business specialization tended to encourage wholesale concentration more than retail. We have no doubt that wholesaling bulked large as a supporting element in American urban growth even before the American Revolution. However, we should not think of this trade in present terms. Even though there were many merchants in Hartford, "yet there was no highly organized or clearly defined wholesale market. Merchants tended to keep on hand a stock to sell either at wholesale or retail. Occasionally they would advertise a special importation of goods to be sold only at wholesale."[38]

Credit Lines Extend into the Back Country. A major factor in

[38] Ibid., p. 13.

establishing and maintaining formal ties was the provision of credit. In most pioneer societies, even as late as the California Gold Rush, cash money tended to be short in supply and barter or quite local paper tended to come into use to facilitate trade. The value of local currency was always in doubt, and proportionately more so because it was more distant in origin or more anonymous in issue. Barter, although more certain in value, was obviously a difficult medium of trade on long-distance undertakings. All these factors encouraged the trading with a known set of participants. "Wholesale merchants helped to finance retailers by selling goods on credit"; thus, when William Moore, a Greenfield, Massachusetts, retailer bought sugar from the leading Hartford merchant, Jeremiah Wadsworth, in July 1787, he was given until September to pay for it.[39] It is probable, therefore, that once lines of trade were established, they showed an endurance quite separate from the conditions of transportation. The opening of railroads in the United States transformed the *possibilities of flow* but seem to have modified the *flow of trade* to a lesser extent, no doubt as much for the functioning of credit as for conditions of movement.

The Ties That Bind Are Traditional. Although money became more plentiful as the American economy developed in the early nineteenth century, particularly after the 1849 Gold Rush, the hold of practice and tradition remained. So long as the retail merchant wished to use another's capital to supplement his own, the ties facilitated the conduct of business. And as standardization of products and specialization in particular lines of trade grew, the retailer had increasing reasons to stick with his customary wholesalers. Much as the introduction of brand labelling strengthened this hold of the wholesaler over the retailer, even before the use of proprietary names, there were advantages to using a consistent source of goods.

HISTORICAL SUPPORT FOR THE AXIOM OF SPECIALIZATION. In a study of the role of middlemen in the growth and shaping of the American economy Fred Mitchell Jones clearly discerns the relationship between *relative location* within the system of trade and extent of product specialization. He tells us that the early

> development of wholesale markets was confined to the large eastern seaboard cities which continued to remain the most important wholesale markets in the United States, but they did not hold their positions unchallenged. The improvements in transportation brought by turnpikes, steamboats, canals, and railways, coupled with industrial progress and an increase in population, caused interior cities to develop as wholesale markets. . . .[40] The development of these markets encouraged retailers to obtain their supplies in these cities rather than in seaboard markets. At the same time, however, industrial progress and improvements in transportation had made it easier for the wholesale merchants in seaboard markets to get supplies. The result was in-

[39] Ibid., p. 156.
[40] Fred Mitchell Jones, *Middlemen in the Domestic Trade of the United States, 1800–1860*. Illinois Studies in the Social Sciences, vol. 21, no. 3 (1937), 14–16.

creased competition, and to meet it merchants and manufacturers made efforts to expand their markets.[41]

The competition between the older coastal entrepôts and the interior wholesaling towns took place along geographical lines. The interior town sought to encourage retail merchants' coming there to shop around for goods in the local wholesaling establishments. The practice they advocated had been typical of the earlier colonial country-trade when the frontier lay only a few miles inland from the initial entrepôts. The nearby town drew the country storekeeper who offered a few basic goods for sale and collected in return therefor the country produce in barter. Both to restore his meagre stock and to realize on the staples he had accepted in payment, it was necessary for that storekeeper to go to the local entrepôt—Boston, Newport, Providence, Hartford, New Haven, New York, and the like—to sell the country produce to those wholesalers who collected staples for export and buy manufactured and tropical products from other wholesalers. Often, of course, the wholesaler who exported staples was the same man who earlier had imported manufactures. In the course of these transactions the country storekeeper came to establish his credit and also to learn the integrity of the wholesaler and his wares. Thus it was not difficult to replace an *actual* journey to the nearby entrepôt with the conduct of business through correspondence, or an infrequent circuit-riding by the wholesaler's agent. As the collecting-and-consuming frontier moved inland, trade probably never began with retailer conflux but commenced, instead, through correspondence and circuit-riding.

The interior wholesaling towns tried to introduce customer conflux, and to a degree they were successful. Perhaps the strongest weapon they held was their ability to employ barter. However, when the coastal merchants began sending traveling agents into the interior on a frequent schedule, they also could begin to assess the realistic worth of local banknotes and commercial paper and to take these in payment for the goods they offered for sale.

The period of credit extended to retailers, particularly to those retailers in the interior, frequently was as long as twelve months. Apparently it was made to coincide with that which the retailers extended to their customers. This in turn depended to a large extent on the crops. The nominal credit, however, was often shorter than this. In reference to the retailers in Cincinnati, it was said in 1817, 'The credit which they receive at Philadelphia is from six to seven months, but they can seldom pay at the specified time, and are then charged 7 per cent interest.'[42]

The stronger capital position of the wholesaler in the coastal entrepôt became his competitive weapon—and in the impoverished interior, it was a compelling one. However to use his position the entrepôt mer-

[41] Ibid., p. 16.
[42] Ibid., p. 17.

chant had often to narrow his range of operations, even as he expanded its field. He might be able to keep tabs from Philadelphia on the value of local bills and wheat and pork in Ohio, but he could hardly deal intelligently in all commodities in all places, particularly at the same time that he was keeping up with the market from which he secured his imports and that for the goods he was selling. In such a demanding complexity of intelligence he would tend to seek the simplicity and security of specialization.

This axiom of specialization represents a critically important addition to, and modification of, the automatic central-place conflux. Any such conflux was most likely to be initial in time and marginal in place. Once a threshold value of trade existed, it was most probable that the coastal merchant would begin to use agency to project competition into the interior. There his economies of scale and superior capital position would often quite successfully destroy the imagined attractions of nearby "central places." Thus, there is strong historical support for the axiom of wholesale location proposed. *If a merchant wishes to specialize, he must broaden his geographical field. If he wishes to narrow his geographical field, he must broaden his lines of trade. And if the tributary area population increases, the wholesaler must increase his product specialization, and thereby improve his service, to retain the same area of trade.*

INSTITUTIONALIZING TIES. What are the implications of all of this to the geography of wholesaling? We have assumed up to now that there are time and distance factors that encourage or discourage the fulfillment of the impulse to trade. Even though this is basically true, we must now add that the facilitation of trade is a matter of institutions, which channel the impulse allowing it to operate with greater ease along one vector than another. For trade probably the most important single consideration after a possible route of trade has been furnished, is *the simple fact that wholesaling works not on hierarchical lines but on historical ones.* Once the ties between the wholesaler and the retailer have been created, they become institutionalized—not on the basis of reducing the intervening distance to the minimum, which is the strong factor in retail trade, but rather on the basis of experience. We may envisage the intrusion of a second wholesaling establishment only in the articulation of a very long distance trade linkage, though even there interception is not too common.

If we may anticipate our story a bit, it might be noted that in the garment industry the ties are from a single center, New York City, to the entire nation, and attempts at interception have had only a very partial success. When, however, New York experienced a public transit strike of some duration in January 1966, and the garment industry suffered the disruption one might expect of the most centrally located of all manufacturing activities, the garment makers were on record as greatly afraid that the customers, once lost, would not return. Thus, the *continuity of trade is a critical factor in wholesaling.*

We end with the conclusion that the interception of wholesale trade, while entirely possible, is not anywhere nearly so likely to take place as is the interception of retail demand by a new and intermediate trading center. And the ability to preserve wholesale trade intact over a large area, even when intervening development takes place, allows that trade to enjoy economies of scale impossible in retailing. That fact goes a long way toward explaining the vigor shown by large old centers even against newer but smaller local wholesaling towns.

CHAPTER 5 *the emergence of the american trading pattern*

Wholesaling was the pioneer form of trade and the strongest force in pushing the mercantile economy into frontier areas. From the foundation of the United States and Canada, local self-sufficiency was viewed as an ideally short-lived stage in the progress toward the better life. Men sought not merely to improve their individual lot by developing virgin land, but also to bring their regional standard of living to a high level for the growing prosperity of their sons. It was not enough to have the material resources of the region so long as these vast countries had even more diverse and lavish resources just awaiting the implementation of trade. But the two nations faced conditions that no European state had ever faced, of seeking to assure a high material culture to people settled thinly on the land. The success that the North Americans had in reconciling these initially incompatible conditions is the main ingredient of the economic wonder of the New World in the nineteenth century. That success was mainly a trading accomplishment that predated the full solution of industrialization on this continent. In a significant way the trading success was founded upon the first efficient and large-scale organization of wholesale trade, which should rank as one of the great achievements of the American nations in the nineteenth century.

The most obvious characteristic of the western trade in America was its almost completely "urban" quality during its early stages. Where there were mining camps with many men, there were stores, often with a surprising array of goods. It is not easy to determine the amount of demand necessary to make a particular good available in these enclaves of urban life scattered about a land still largely supported by the harvest of raw resources. The willingness of frontiersmen to pay high prices for goods meant that the impulse to trade was satisfied with a smaller accumulation of demand than would have been the case in a European economy. We make a mistake if we think that there had to be a vast demand to create trade. Christaller, and most of his disciples, work within a narrow present-day frame that tends to hold that the threshold for a trade is predicated on the minimum support necessary to keep going a shop specializing in that line. Evidence cited elsewhere [1] supports the rejoinder that before such specialization took place, the threshold was not the one that would operate for a specialty shop but rather the one that formed the floor of trade that must rest under the operations of a general merchant. Through combining the trade in a number of lines, local merchants could secure an over-all living, not one based on excessively high prices for a narrow range of items.

WHOLESALING BUILDS THE CITIES OF THE FRONTIER. Several important conditions flow from the simple fact that in America there was a relatively large and continuous demand for those goods that lent themselves to wholesale trading. In the Middle Ages the demand for specialized products had been met for the mass of the people by an increase in the numbers of local artisans. Only the tiny aristocratic group could command skill from elsewhere, and they often imported the workman rather than his product. But with the settlement of North America, both in the United States and Canada, trade was natural and soon became backed up by a tradition of the relatively large middle classes of petty merchants, professional men, small manufacturers, and prosperous freehold farmers buying the product of many skills. Unlike the feudal lord, they had to import the product rather than the workman, so it came about that the traders expanded in numbers and importance much as the artisans had in the medieval times. America had traditional craftsmen, but they were unusual and tended to be found only in the largest and wealthiest port cities. John Hull, Benjamin Goddard, and Paul Revere can serve as examples; but we should not think they represent a vast group and they were even more common in colonial times than later. Much larger and more characteristic of the American economy were the merchants who dispatched goods to the interior and often settled collateral family branches near the frontier to make sure the trade worked efficiently and effectively.

The numbers of wholesale merchants in Hartford in colonial times have already been cited. They did not die as a class with independence, rather the opening of the West encouraged an occupational expansion. The colonial towns—Boston, Newport, Providence, Hartford, New Haven, New York, Philadelphia, Baltimore, and Charleston—were traders' towns built at what we might call the *unraveling point of trade,* that place which represented the last point common to a sufficient number of *supply lines* to the frontier to assure a reasonable institutional income to the trader. Onward from the unraveling point, only *strands* connected

[1] James E. Vance, Jr., "Emerging Patterns of American Commercial Structure," Papers for the Lund Symposium on Urban Geography, Lund Studies in Geography, Ser. B Human Geography No. 24, 1962 (Lund, Sweden), 485–518.

to the customer; and none was strong enough to impel trade across the Atlantic. Only rearward from the unraveling point was the flow of a sufficient force to create periodically successful trade impulses.

THE UNRAVELING POINT. This situation is quite different from the one envisaged under central-place theory. We are not talking about customer gravitation but rather about trade flow, which is much more easily rationalized and organized than is retail gravitation. Additionally, the unraveling point is not quite the same thing as the threshold. We can assume that the threshold must have a relatively low unit value, otherwise there could not be a sufficient number of centers of a particular type of trade to permit all people reasonable access to the goods. There is no such leveling force working on the location and number of unraveling points. *The individual unraveling points will be as far forward as necessary to command a satisfactory market, but merchants located in each can counter moves for further advance* by quantity price adjustments. *In wholesale trade the flow is dismembered but not the trading place, as in retail trade.*

The ability of the wholesale town to protect its traditional position through price variation is not absolute. Dominance of the trading area comes from the carrying out of a dual function: the provision of place utility for goods and the anticipation of future markets located at a distance from the source of goods. The second of these functions requires trading establishments to follow the frontier, even if slowly and none too closely.

The fact that the appraisal of the market forced traders inland does not contravene the notion of the unraveling point. It meant that Cincinnati lay over the mountains from the main sources of manufactured products; it did not mean that there were many Cincinnatis. There were some others, however, and they were determined in location almost entirely by the transportation technology that ruled at the time they were founded. Cincinnati came to dominate at first because the flow was across the part of the mountain barrier that stretched between Cumberland Gap and central Pennsylvania. In the early trans-Appalachian West that was where the men lived.

With the spread of the economic frontier farther into the Middle West, into the areas north of the Ohio valley, it was hard for Cincinnati to serve alone as an unraveling point. Particularly after 1825, when the Erie Canal opened, trade tended to flow along the easy path of the canal to Buffalo (Black Rock). Thence lake boats carried the goods to lake ports and these became, and remain, important trading emporia of the Middle West. Coasting along the American shore of Lake Erie, the first place that a good unraveling point was to be found was at the mouth of the Cuyahoga where Cleveland was established as a trading port, later connected inland by the Ohio and Erie Canal (finished to Portsmouth on the Ohio River in 1830). Toledo was at the mouth of the Maumee and was in turn connected by canal to Cincinnati (in 1846) and Evansville (in 1851). Detroit already had been singled out by the

French, and all the Americans did was to build a number of roads leading along the possible vectors of trade toward the interior.

But Cleveland, Toledo, and Detroit never quite matched Cincinnati in relative importance as an unraveling point. On Lake Michigan, Green Bay had also been selected by the French, but its hinterland was rather slow in developing, whereas that behind the sand dunes at the mouth of the Chicago River grew more rapidly. It was at that place that the farthest interior major unraveling point came into existence and throve as a lake and canal port for a generation before it became the *summa* for the railroad age.

THE UNRAVELING POINT IS THE METROPOLIS. The weakness of settlement stratification in the early decades of American expansion is a fact that flies in the face of most of our thinking on the origin and support of cities. The argument is normally advanced that towns grow up to care for central-place functions—mainly retailing and services except in the largest places—and that these places mesh into a stratified network of towns with importance that increases with intercity spacing. The central-place model leads us to expect (1) a large number of closely spaced minor trade centers providing everyday essentials, (2) a somewhat reduced array of larger places providing less frequently desired goods, and (3) other classes of towns of increasing size and trading complexity but ever decreasing numbers. This is the hierarchical structure with its view of the carefully nested functional pattern.

What evidence can be brought together on American trade and trading patterns in the middle of the last century does not conform to this postulated pattern. Rather we find such a great simplification of the nesting principle that we have two, or at the most, three levels of trading places with great scalar difference between them. The *fundamental trading center* is small and simple, possessing one or two commodity-combining general stores, which have the normal threshold value for the establishment, but a submarginal value for the trade in any particular class of goods. This commodity combining allows the conduct of business in any single line of trade in the most dispersed fashion that can be imagined. In addition, these basic towns have simple service functions, notably the offices of attorneys, physicians, and dentists. The underpinnings of a net of basic towns should, according to theory, be capped by ever-reducing courses of more elaborate places. But it is this superstructure that we do not find. Table 3, drawn from a directory of businesses in Illinois and Missouri in 1854, the year the first railroad reached Chicago, shows the tendency in this area toward the development of a two level city structure, comprised of basic towns and large merchant cities. These latter, it is proposed, can be better explained with respect to location and numbers through the notion of the unraveling point than through the mechanism of hierarchical assembly of parochial trading areas.

If we accept the idea that in a *pioneering* economy the demands placed on the wholesaling structure are relatively greater than those

Table 3. Types of Business Found in Illinois and Missouri Towns (1854)

(Est. analyzed)	*Express Agents*	*Freight Agents*	*Auction & Commission Merchants*	*Banks*	*Books & Stationery*	*Boot & Shoe Dealers*	*China, Glass & Queensware Dealers*	*Clothing Merchants*	*Commission & Forwarding Merchants*	*Dry Goods*	*Furniture Dealers & Manufacturers*	*Groceries*	*Hardware, Etc.*	*Hotels*	*Printers*	*Produce Dealers*	*Tobacco Dealers*
ILLINOIS																	
Chicago (598)	1	4	15	17	8	22	3	42	48	36	37	222	26	38	15	26	29
Galena (145)	1	–	2	3	2	16	–	15	9	23	5	46	5	8	2	6	2
Peoria (100)	–	–	–	4	3	5	–	11	–	35	4	18	10	8	2	–	–
Rock Island (56)	1	–	–	2	1	2	–	4	2	11	3	17	5	5	1	2	–
Springfield (69)	–	–	2	3	2	1	–	–	1	23	7	17	4	5	2	–	2
Illinois Total (968)	3	4	19	29	16	46	3	72	60	128	56	320	50	64	22	34	33
MISSOURI																	
Booneville (8)	–	–	–	1	1	–	–	–	–	–	–	6	–	–	–	–	–
Columbia (6)	–	–	–	–	–	4	–	–	–	–	–	2	–	–	–	–	–
Hannibal (42)	–	–	–	1	2	12	–	5	3	–	–	10	–	5	–	–	4
Independence (5)	–	–	–	–	–	–	–	5	–	–	–	–	–	–	–	–	–
Jefferson City (6)	–	–	–	–	1	–	–	–	–	–	–	–	–	5	–	–	–
St. Joseph (11)	–	–	–	–	1	3	–	–	–	–	–	3	–	4	–	–	–
St. Louis (1171)	3	15	91	17	25	120	14	122	–	108	50	345	19	15	10	–	108
Distinguished as wholesale	–	–	–	–	–	14	–	–	–	23	–	58	14	–	–	–	–
West Port (4) (later Kas. Cty.)	–	–	–	–	–	2	–	–	–	–	–	–	–	2	–	–	–
Missouri Total	3	15	91	19	30	155	14	132	3	131	50	424	33	31	10	–	112

Source: *Montague's Illinois and Missouri State Directory for 1854–55* (Principal Men of Business in the States of Illinois and Missouri . . . Complete Business Directory of City of St. Louis). St. Louis: 1854.

exactions in more elaborated economic systems, then the strange disappearance of the "middle class of towns" is more easily understood. We have already seen that commodity combination, in general stores, is joined with the frontier acceptance of high transport charges to allow the distribution of manufactures over very wide areas. The system of country general stores was sufficiently flexible to offer the patient frontiersman the ultimate satisfaction of a great variety of wants. In such a context, the involved hierarchical arrangements of central-place theory did not emerge in America until much later, if as yet they have. And as the explanation of the *processes* of growth in new land economies, central-place theory is too weak to survive.

HOW DID TRADE FUNCTION WITHOUT CENTRAL PLACES? THE ENTREPÔT CITY. Without the constant flow of customers to market towns trade was accounted for by three geographical elements. There were the major cities that served as the unraveling points of major channels of supply.[2] These we might best call the *entrepôt cities* from which thin and normally *periodic threads* of trade connected (1) to *workplace towns* where men harvested a wealth from untended resources, or (2) to *fundamental trading towns* in the pioneer farming areas.

Beyond the Missouri, urban geography was first that of workplace towns such as Denver City (1859), Cheyenne (1867), or Virginia City (Nevada in 1859, Montana in 1863). The location of these places had no hierarchical basis. Denver, although it grew to be a great emporium, was set up to serve miners who sought placer gold in Cherry Creek and later traded that gold for staples. Denver, continuing as the trading town after the gold search led to the richer gravels around Central City and even farther away, was soon turned into an even more interior unraveling point than St. Louis or Chicago. The transformation was common in the West. Cheyenne was a railroad town in 1867 with no tributary area at all save in the very specific context of the line westward to Ogden. As settlement came, in a thin scatter of relatively self-sufficient ranches on the High Plains and in the Laramie and other high basins, Cheyenne became an entrepôt city of modest attainments.[3] But again it lacked the stratified connections that central-place theory would lead us to expect of it.

Thus, for staple foods, work clothing, whiskey, and other ordinary needs, Wyoming ranchers probably turned to Cheyenne. But for the more specialized demands they would have looked to Omaha, where the Union Pacific Railroad brought a still greater number of threads together, or to Chicago, where the greatest number of frontier demands could be collected. The result was a *wholesale stratification* with the

[2] See David Revzan, *Wholesaling in Marketing Organization* (New York: Wiley, 1961) for the original and authoritative statement on wholesale trade as a channel of supply.

[3] For a more complete discussion of this transformation see the author's work on the building of the railroad and its impact, "The Oregon Trail and Union Pacific Railroad, A Contrast in Purpose," *Annals of the Association of American Geographers*, vol. 51 (1961): 357–79.

rancher, or his urban peer, the small-town general merchant, turning to successively more distant places to satisfy more unusual needs. In most cases the purchaser did not go to the trading establishment, at least after setting up initial business relations, so the impact of distance was very different from that envisaged under Christaller's conception of retail trade nesting.

The Early Entrepôt Cities of America. Argument along these lines requires clear sustaining evidence. The basic thesis holds that the earliest support for enlarged cities, beyond the encouragement offered

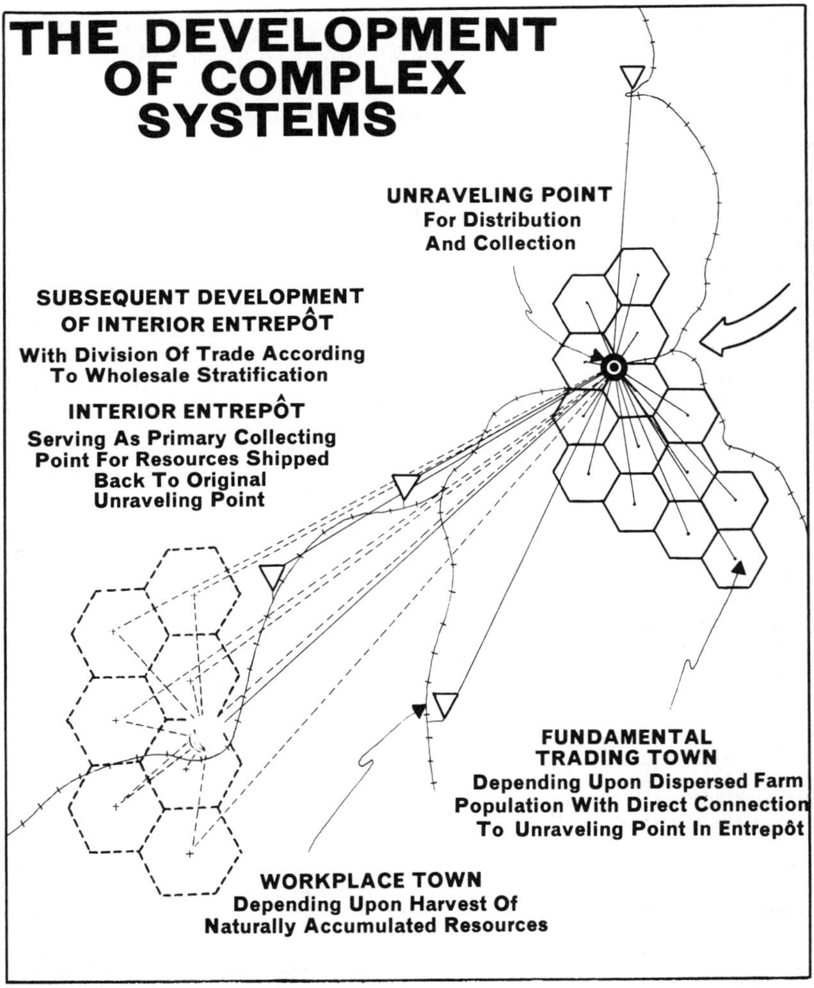

Figure 7

by the population in the close vicinity, comes from wholesaling rather than retail trade. In this view only a single entrepôt in each major region will rise to large city size on the basis of wholesale trading, although different supports may work to enlarge other cities in the region. Since data on wholesale trade were first collected in 1930, there is no direct statistical evidence to test this proposition. Fortunately, there are some related statistics that were collected much earlier.

In Table 4 the census date when each city was recorded with a population of 5,000 and 50,000 is shown. This information serves as a rough approximation of the time when the city might class as an important population concentration (the 5,000-population date) and when it reached the semimetropolitan status characteristic of the entrepôt (the 50,000-population year). The meaning of the figures is greatest in the earlier censuses and least in the most modern. Thus, a place of 5,000 people in 1790 was probably "metropolitan" in regional importance if not in numbers and internal complexity.

When the first census was taken in 1790 there were already eight places with 5,000 people or more but only two, New York and Philadelphia, with more than 25,000. None reached 50,000 until 1800 when New York (present composition) reached that size. The original eight towns of 5,000 were trading places engaged in maritime activities, and during the next twenty years fifteen more towns of 5,000 were added, but with no more than two, Schenectady and Lancaster, lying beyond the reach of the tide. In 1820 census six towns joined this string of ports in the 5,000 population class, but none of the additions was a true port. Troy and Hudson were on the Hudson River, but they were satellites of Albany rather than ports in their own right. The other four towns added in 1820 were, interestingly, the entrepôts of the new West—Pittsburgh and Cincinnati on the Ohio, and Lexington to the south and Columbus to the north of the river. In the next thirty years 85 other towns in the United States reached a population of 5,000 but only fourteen were in the Old West made up of Kentucky, Ohio, or Indiana; even then, all but six were Ohio River or Great Lakes ports. Only Indianapolis was not on these waterways or a canal. In 1850 these three states had 17 per cent of the population of the United States and nearly 16 per cent of the towns of 5,000. Despite a relatively regular terrain and consistent land settlement pattern, the towns that reached 5,000 were pretty much along the lines of long-distance trade.

Turning back to the Atlantic seaboard, we find the emergence of workplace towns in the 1830 census. Lowell, Lynn, Springfield, and Taunton in Massachusetts, Dover in New Hampshire, and Warwick in Rhode Island were five of the fifteen towns to reach 5,000, and they were clearly industrial places. The impact of the Erie Canal in setting up potential distributing towns was obvious in the case of Utica, Rochester, and Buffalo. Other emergent towns in 1830 were either metropolitan satellites (Newark and Cambridge) or new entrepôts in the West (Nashville and Louisville). Reading was both a trade center and a manufacturing town. In 1840 twenty towns were added to the 5,000-population group, of which four were factory towns in the Northeast (Fall River,

Table 4. Census in Which American Cities First Were Recorded with Populations of 5,000, 25,000, and 50,000

Census	*50,000*	*25,000*	*5,000*	
1790	—	New York Philadelphia	Baltimore Boston Charleston Gloucester	Newport Providence Salem
1800	New York	Baltimore	Albany Newburyport Norfolk Portsmouth, N.H.	Richmond Savannah Schenectady
1810	Philadelphia	Boston	Alexandria Lancaster New Bedford New Haven	New Orleans Petersburg Washington
1820	Baltimore	New Orleans	Cincinnati Columbus, O. Hudson	Lexington, Ky. Pittsburgh Troy
1830	Boston	Charleston	Buffalo Cambridge Dover, N.H. Hartford Louisville Lowell Lynn Nashville	Newark Reading Rochester Springfield, Mass. Taunton Utica Warwick, R.I.
1840	New Orleans	Albany Cincinnati Pittsburgh	Auburn, N.Y. Augusta, Ga. Augusta, Me. Cleveland Dayton Detroit Fall River Frederick Harrisburg Lynchburg	Mobile Nashua New London Paterson Portsmouth, Va. St. Louis Wheeling Wilmington, Del. Wilmington, N.C. Worcester
1850	Albany Cincinnati Pittsburgh St. Louis	Buffalo Chicago Louisville Lowell Newark Providence Richmond Rochester St. Louis San Francisco Troy Washington	Beverly Biddeford Bridgeport Burlington, Vt. Camden, N.J. Chelsea, Mass. Chicago Chicopee Chillicothe, O. Columbia, S.C. Columbus, Ga. Concord, N.H. Covington Easton Elizabeth Erie	Montgomery New Albany, Ind. Newport, Ky. Newton, Mass. Norristown Northampton Norwich, Conn. Oswego Peoria Pittsfield Pottsville Quincy, Ill. Quincy, Mass. Racine Saco Sacramento

Table 4 (continued)

Census	*50,000*	*25,000*	*5,000*	
1850			Fitchburg Haverhill Indianapolis Jersey City Lafayette, Ind. Lawrence, Mass. Macon Manchester Memphis Milwaukee	Sandusky San Francisco Springfield, O. Steubenville Poughkeepsie Syracuse Trenton Watervliet Zanesville
1860	Buffalo Chicago Louisville Newark Providence San Francisco Washington	Cambridge Cleveland Detroit Hartford Jersey City Milwaukee Mobile New Haven Portland, Me. Syracuse	[Too numerous to detail]	
1870	Cleveland Detroit Jersey City Milwaukee New Haven Richmond Rochester	Columbus, O. Dayton Fall River Indianapolis Lawrence Lynn Memphis Nashville Paterson	*(25,000 continued)* Kansas City Reading Savannah Scranton Springfield, Mass. Utica Wilmington, Del. Worcester	
1880	Cambridge Columbus, O. Indianapolis Kansas City Lowell Paterson St. Joseph Syracuse Toledo Troy Worcester	Atlanta Bridgeport Camden, N.J. Covington Denver Elizabeth Erie Evansville Fort Wayne Grand Rapids Harrisburg	Hoboken Lancaster Manchester Minneapolis New Bedford Oakland Omaha Peoria	Quincy, Ill. St. Joseph St. Paul Salem Terre Haute Trenton Wheeling

Nashua, Paterson, and Worcester), three were slow-growing Atlantic ports (New London, Portsmouth, Va., and Wilmington, N.C.) and nine were local trade centers or Ohio River ports. But also in 1840 the farther-out entrepôts began to show up in the rise of Cleveland, Detroit, and St. Louis.

By 1850 the towns of 5,000 were relatively less important than they had been, and the impact of rapid industrialization in the Northeast and eastern Middle West was matched by the emergence of many local trading centers in the older settled regions. There were 51 places added in 1850, of which number perhaps half could class as workplace towns and the other half as local trading centers for smaller regions. The next band of entrepôts shows up with Chicago, Memphis, and San Francisco. By 1850 the tide of trading places had nearly reached the Mississippi, but only as a gentle ripple. And west of the river there were only three towns of 5,000—St. Louis, San Francisco, and Sacramento. Each epitomized wholesaling.

If we look upon the city of 50,000 as a true distribution center and propose that from these places the first unraveling of trade begins, the pattern shown in the United States should support that view. The first five censuses of the nineteenth century added one city of 50,000 each decade. The first four were the great colonial and Federal period ports —New York, Philadelphia, Baltimore, and Boston—and the fifth, New Orleans, was the first great port of the West. In 1850 the major western river entrepôts—Pittsburgh, Cincinnati, and St. Louis—provided most of the additions to the group of large cities. Only Albany, which had a rather definite entrepôt function from its early status as a major inland town and the port for the Erie Canal after 1825, grew apace with the more interior distributing centers. In 1860 the pace increased to seven new cities in a decade, but still a majority were in the West, which remained unsettled in many parts and largely underdeveloped in comparison with the already industrializing and stratifying Northeast. Buffalo, Chicago, Louisville, and San Francisco were the new large cities, and their rise meant that there were eight cities of 50,000 east of the Appalachians and eight to the west, and none south of the Potomac except for New Orleans far to the west. This would make a strange central-place structure.

When we look back at the group of towns (5,000 population or more) we find only 34 out of 116 (29 per cent) were west of the Appalachians in 1850, at the same time when four of the nine large cities were in the West. In 1860 the score began to level, with just over half the total number of towns of 5,000 lying west of the Appalachians. Even so there were only twelve towns of 5,000 west of the Mississippi in 1860, when there were 154 east of it; but two of those twelve towns beyond the river reached the size of major cities of 50,000—St. Louis and San Francisco.

It would not advance our story to expand this appraisal of the growth of American cities. The evidence supports the view that in both the colonial and Federal economy on the Atlantic seaboard and the pioneering economy of the West before the Civil War there were a few predominant distributing points and entrepôt cities, and a stunted development of lesser trading places. Ultimately those smaller places did develop, chiefly as an outgrowth of industrialization and dense commercial farming. Even now in the open lands of the Great Plains and the Rocky Mountain West, hierarchy is not clearly evident and a few disproportionately large towns serve the population.

Wholesaling and the Entrepôt City. No direct measure of whole-

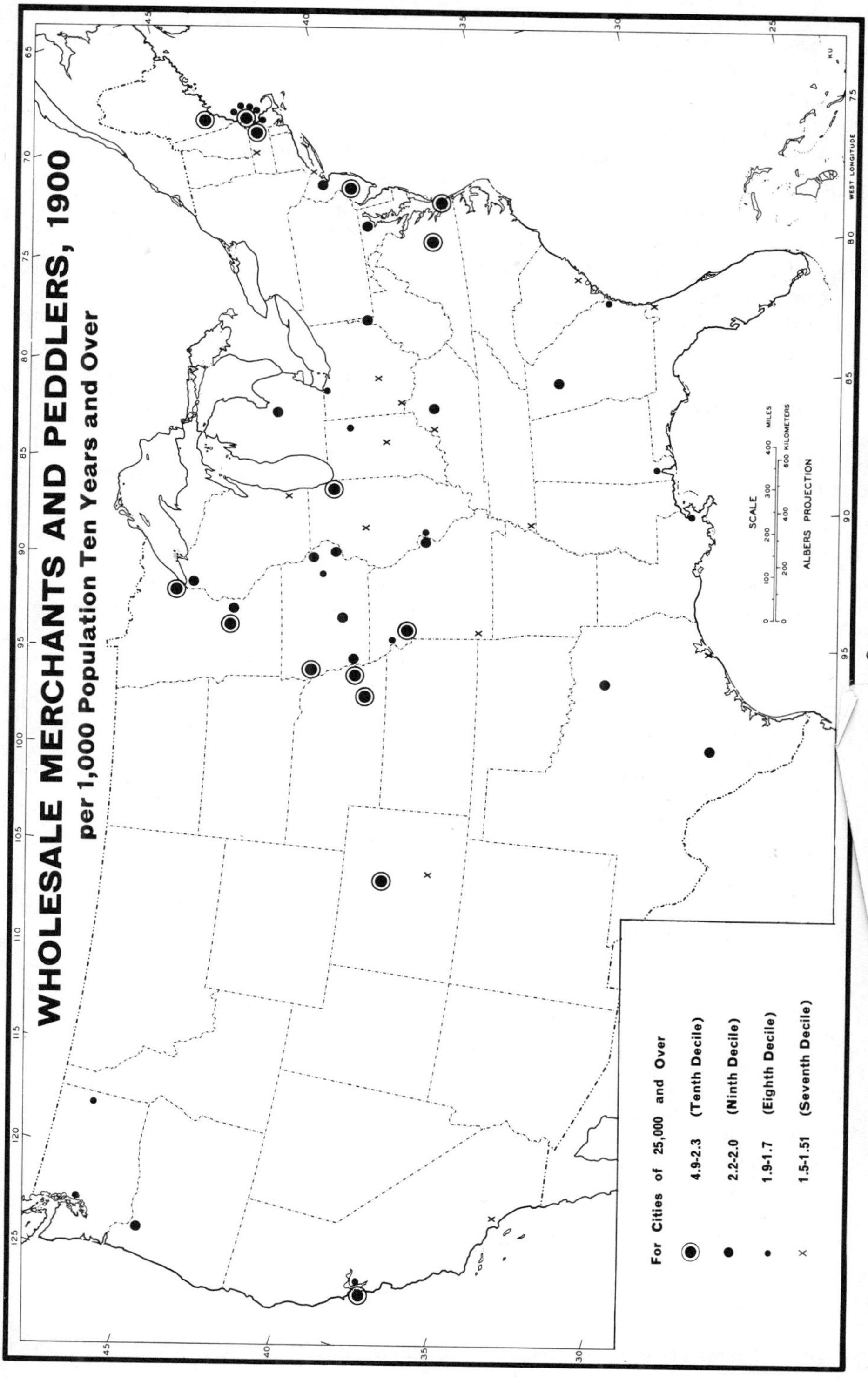
WHOLESALE MERCHANTS AND PEDDLERS, 1900
per 1,000 Population Ten Years and Over
For Cities of 25,000 and Over
4.9-2.3 (Tenth Decile)
2.2-2.0 (Ninth Decile)
1.9-1.7 (Eighth Decile)
1.5-1.51 (Seventh Decile)
SCALE
0 100 200 300 400 MILES
0 200 400 600 KILOMETERS
ALBERS PROJECTION
WEST LONGITUDE
KU

re 8

sale trade in cities is possible until 1929, but there is indirect evidence that we may use, in the tabulation of Wholesale Merchants and Dealers, which began to be published in the occupation tables starting with the 1900 census. In Table 5 the first 30 cities, ranked by total number of Merchants and Dealers, are shown. These figures are rough; in the absence of specific sales data this limitation must be accepted. Nonetheless, certain rather illuminating conditions are brought out when we compare wholesale and population rank. Those cities that show up high in wholesaling (when compared to population) are mainly the entrepôt cities still on the outer edges of the ecumene of 1900. Richmond, Kansas City, Omaha, Portland, Oreg., Minneapolis, Denver, St. Paul, Los Angeles, and San Francisco were then beyond the edge of the manufacturing belt and outside the corn belt or at its margin. Thus, the two forces possibly creating central-place hierarchies were absent, but the force of wholesale trade

Table 5. Cities of the U.S. Ranked by Number of Merchants and Dealers

	City		
Rank	1900	1910	1920
1	New York City	New York City	New York City
2	Chicago	Chicago	Chicago
3	Philadelphia	Philadelphia	Philadelphia
4	St. Louis	Boston	San Francisco
5	Baltimore	Baltimore	St. Louis
6	San Francisco	St. Louis	Baltimore
7	Kansas City, Mo.	San Francisco	Boston
8	Boston	Pittsburgh	Los Angeles
9	Minneapolis	Cincinnati	Detroit
10	Cincinnati	Los Angeles	Seattle
11	Milwaukee	New Orleans	Pittsburgh
12	Detroit	Detroit	Minneapolis
13	Buffalo	Kansas City, Mo.	Kansas City, Mo.
14	New Orleans	Cleveland	New Orleans
15	Louisville	Newark	Cincinnati
16	Cleveland	Milwaukee	Cleveland
17	St. Paul	Buffalo	Milwaukee
18	Denver	Minneapolis	Buffalo
19	Newark	Louisville	Portland, Oreg.
20	Omaha	St. Paul	St. Paul
21	Indianapolis	Denver	Newark
22	Pittsburgh	Seattle	Louisville
23	Jersey City	Portland, Oreg.	Oakland
24	Columbus, O.	Jersey City	Denver
25	Washington	Columbus, O.	Indianapolis
26	Somerville, Mass.	Atlanta	Jersey City
27	Richmond, Va.	Memphis	Omaha
28	Portland, Oreg.	Washington	Memphis
29	Toledo	Indianapolis	Dallas
30	Los Angeles	Toledo	Rochester

Source: Occupation Tables, 12th, 13th, and 14th Census.

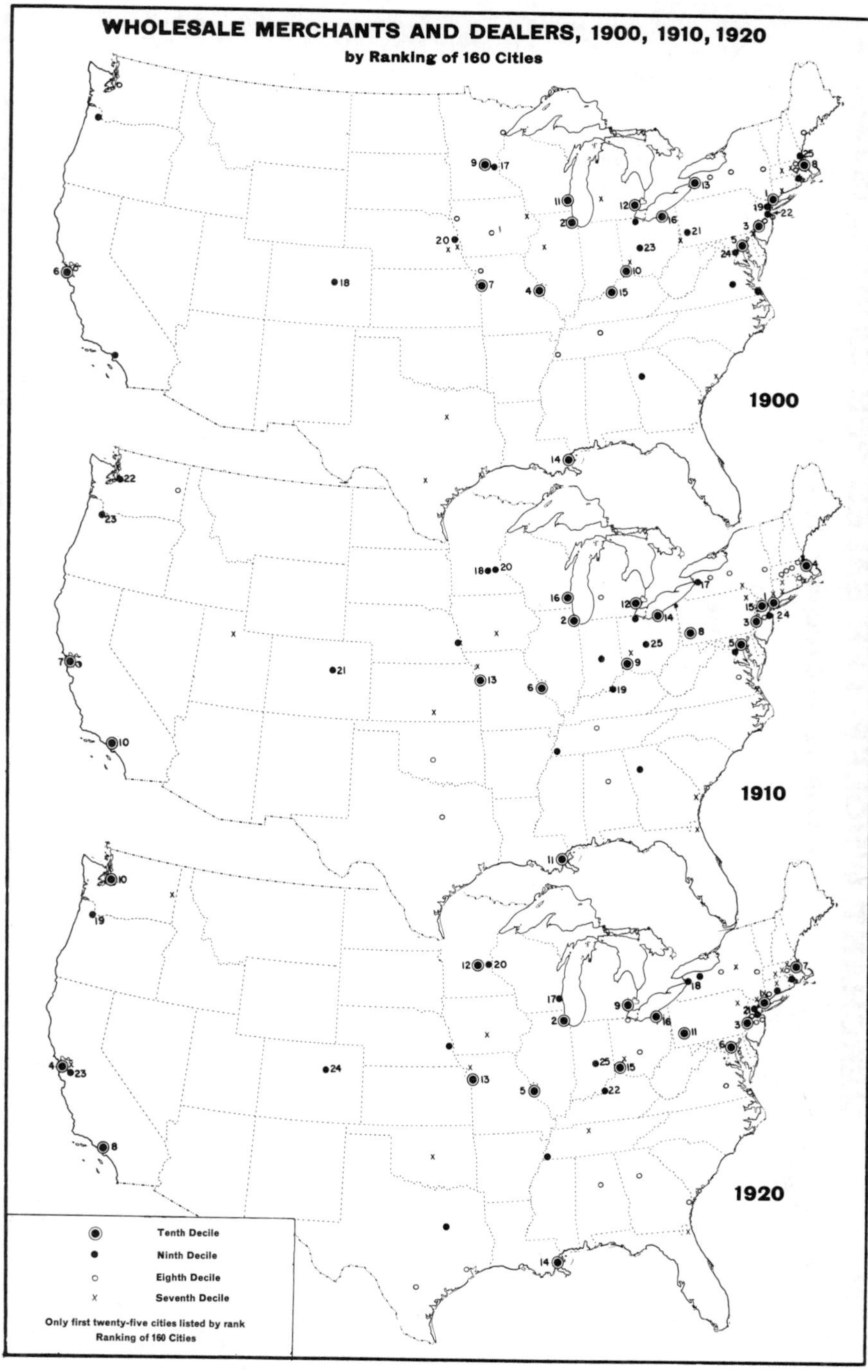
WHOLESALE MERCHANTS AND DEALERS, 1900, 1910, 1920
by Ranking of 160 Cities
1900
1910
1920
Tenth Decile
Ninth Decile
Eighth Decile
Seventh Decile
Only first twenty-five cities listed by rank
Ranking of 160 Cities

Figure 9

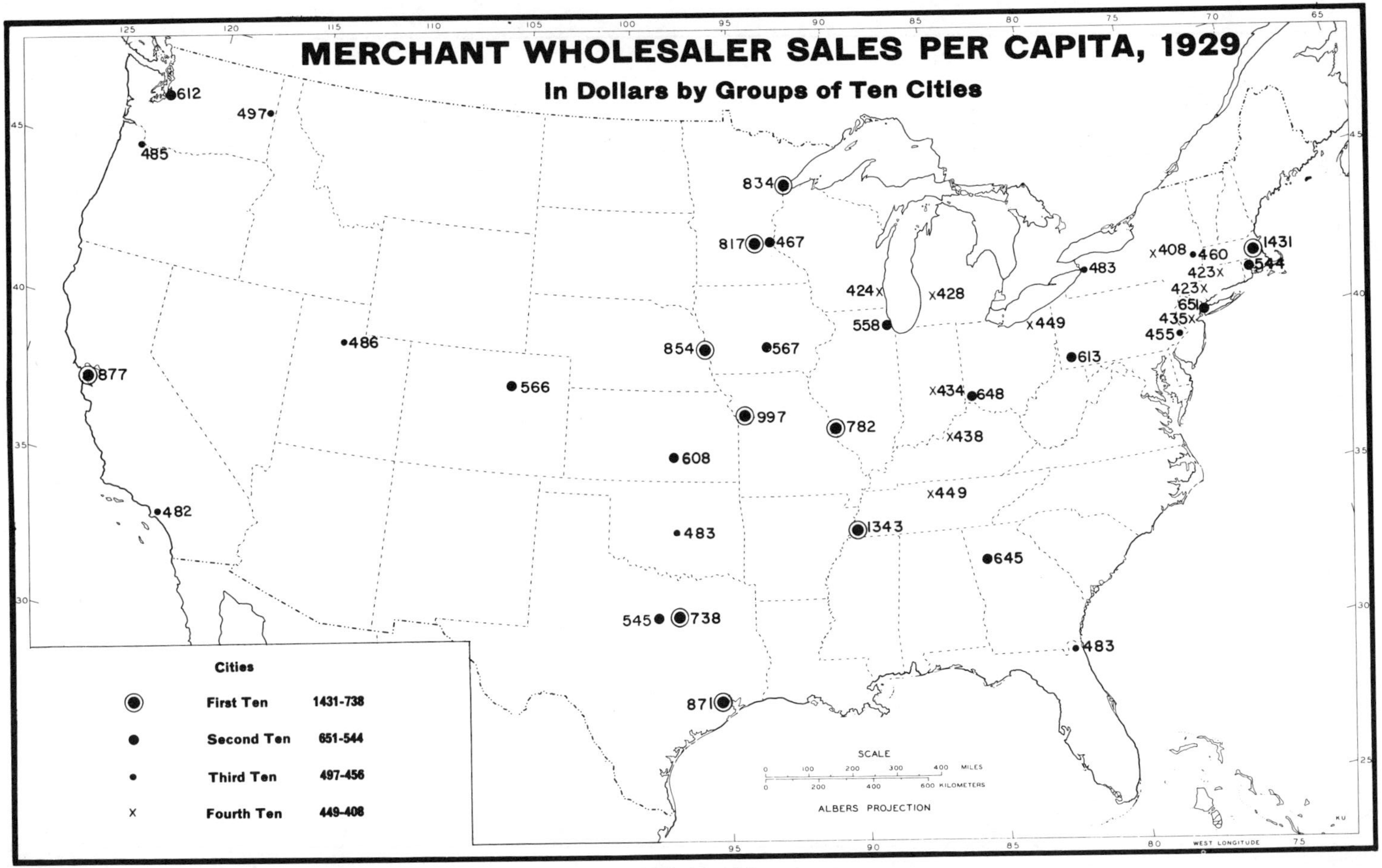
MERCHANT WHOLESALER SALES PER CAPITA, 1929
In Dollars by Groups of Ten Cities
612
497
485
834
817
467
1431
408
460
544
423
423
651
435
455
483
424
428
558
449
613
486
854
567
877
566
434
648
997
782
438
608
449
482
483
1343
645
545
738
483
871
Cities
First Ten 1431-738
Second Ten 651-544
Third Ten 497-456
Fourth Ten 449-408
SCALE
0 100 200 300 400 MILES
0 200 400 600 KILOMETERS
ALBERS PROJECTION
WEST LONGITUDE
KU

Figure 10

was not. All the cities showing a parity between population and wholesaling or a lesser position in wholesaling than in population were in the manufacturing and corn belts. Other evidence suggests that part of this poor showing for some of the older cities, such as Boston and Pittsburgh, stems from the existence, already in 1900, of a considerable amount of wholesaling in outlying metropolitan communities. The inclusion of satellitic wholesale centers would merely raise the two cities to the level of parity enjoyed by Philadelphia and New York, which had annexed most of their immediate neighbors in 1854 and 1899, respectively.

A more precise comparison of wholesaling and population is possible in 1930 because sales data for 1929 were collected along with the population data in 1930. If we rank cities by population and wholesale sales, we can again measure the divergence between the two rankings. As we found in 1900, the cities away from the manufacturing belt and the corn belt show the strongest development as entrepôts. There are a few anomalies. Peoria ranks very high in wholesaling and lies right within the corn belt and Wilmington, Del., has the same position but within the manufacturing belt. Nonetheless, of the 25 highest cities, all are in areas where the possibility of a central-place hierarchy is limited, either because the cities lie in the areas of developing regional economies, as in the West and South, or at the edge of the corn and manufacturing belts. Of those 25 cities wherein wholesaling ranks highest in relation to population rank, only Boston, Wilmington, and Peoria might be thought to lie within the manufacturing or corn belts. Albany, Utica, or Evansville stand at their boundaries.

In 1930 the cities that showed the best balance between wholesaling and population rank were the large traditional regional centers—New York, Chicago, Minneapolis, Denver, Nashville, Pittsburgh, Cincinnati, Birmingham, Philadelphia, Seattle, St. Louis, Detroit, Buffalo, Portland, Oreg., and Syracuse—which made up half the places in which the two rankings were on a par or within ± three ranks of each other. The cities with great negative disparity between wholesaling and population rank tended to be industrial towns—such as Flint, Jersey City, or Worcester—satellitic cities—such as Long Beach, Camden, or Oakland—or cities clearly integrated into well-developed regional systems as subdominant trading centers—such as Milwaukee, Baltimore, Columbus, or Fort Wayne.

Thus, even as late as the time of the first census of business, the reality of the pioneering function of the wholesaling city could be seen. Beyond the edge of the true ecumene of the United States, the entrepôt city assumed an importance in trade far above its population significance in the nation as a whole.

LATER SUPPORT FOR THE NOTION OF THE WHOLESALING ENTREPÔT. Seeking to measure "centrality" of cities led Siddall in 1961 to plot wholesaling–retailing ratios for the 56 Standard Metropolitan Areas that had 300,000 or more people in 1950. In addition he computed the Quantitative Measure of Centrality furnished by determining the cities that had a disproportionately large employment in wholesale trade. In 1950, 18.2 per cent of all workers in trade in the United States were engaged in what was classed as wholesaling. This Siddall used as his norm for wholesaling activity in individual cities. Putting aside arguments that must ultimately be raised about the validity of using the national average as a city norm, it is possible in this way to identify those cities where the number of wholesale workers is proportionately greatest. There should be no surprise that Siddall found much the same pattern that was appearing in 1900 and was easily discernible in 1930. The cities of greatest relative employment in wholesaling were those outside the corn and manufacturing belts, or those that lay on the border of either belt.[4]

The explanation that Siddall advances for this pattern is related to his basic assumption that cities in the United States live in a hierarchy structured by central-place theory.

As we shall see in the remainder of this chapter, the pattern of wholesale trade conforms not to the central-place stratification but rather to distinct *regional structures,* which are the physical expression of historical–geographical forces. But unlike the concept of regions and regionalism, these regional structures are not idiographic. Their nature is explained by a few general dynamics that have application to all regions and times basically because they are fully able to comprehend changes wrought by time.

Regional Structures of Wholesale Trade

To understand the several regional structures that characterize wholesale trade in Anglo-America, and possibly in other well-developed capitalist economies as well, we must begin with the notion of the *unraveling point* already discussed. That point interests us both with respect to its characteristic location and its numbers relative to the regional development. The two facets are related in a straightforward and obvious manner. Great numbers of unraveling points will tend to produce a dense pattern of such cities, and few unraveling points within a major region will quite obviously have to be well separated. These are the limits of relationship but not their explanation.

WHOLESALING IN A PIONEER REGION. There is only one major region of the United States, Alaska, where we still find pioneering in the economic and social sense, but Canada is well provided with examples. Thus, it is not exclusively a historical interest that causes us to ask, What is the trading pattern that characterizes pioneering economies?

Harking back to the environmental conditions that shape trade, we can see that in a pioneering stage the settlement pattern will be thin, and disposable income will be modest but concentrated in use at the two poles of demand, prime staples and true luxuries. The main support for cities will come from trading, either in the sense of securing and distributing manufactures or in collecting and forwarding the natural

[4] William R. Siddall, "Wholesale–Retail Trade Ratios as Indices of Urban Centrality," *Economic Geography,* 37 (1961): 124–32.

produce—mineral, animal, or forest. Cities will be few in number and widely separated. Even more important, they will be located at the margin of the region they serve and on the side toward the source of manufactures and the destination of the natural produce that has been harvested. Thus, quite in contrast to the assumed central-place structure (with a net of towns thrown rather uniformly across the land in relation to retail trade volumes) the *entrepôt alignment* is much more typical of pioneering economies. Because relatively long demand accumulation is characteristic of pioneering economies, as we discovered in the last chapter, it is possible for wholesale cities to be located at some distance from the ultimate customer. The specific location of these unraveling points will depend largely on the particular type of transportation in use.

Applying these ideas to real conditions, we find that this notion of the entrepôt alignment provides us with a far more encompassing explanation of the early urban geography of Anglo-America than would central-place theory. The statistics on city growth (Table 4) show clearly that the first wholesaling towns, the first towns in fact, had just such an entrepôt alignment and that for several decades urban growth came not in those interior towns that might have been used to shape a hierarchy —but rather in the smaller members of the original alignment. So long as sea transport was the main means of movement, either to and from Europe or among the colonies and original states, the alignment was situated in a string of ports, which might be dotted along the coast, as in New England where the rivers were short and barred by falls a short distance inland; or the alignment might extend in a linear fashion up the course of a navigable river, such as the St. Lawrence, Hudson, or Delaware. Along the Connecticut River towns were late in coming into wholesaling, despite very early initial settlement, because the interruption of navigation at Enfield was not overcome until relatively late, by which time the earlier entrepôts at Albany and Montréal had captured most of the back-country. The persistence of early trading patterns (because of traditional wholesale dealing and normal demand accumulation) allowed the frontrunners to hold that lead.

It was not the possible central places on the coastal plain that became the next entrepôts as the frontier moved inland, it was the river and lake alignments in the Middle West. Pittsburgh, Cincinnati, St. Louis, Buffalo, Cleveland, Detroit, and Chicago were the next recruits. Lexington, Ky., and Columbus, Ohio, became modest towns but not full entrepôts. Still later as the frontier moved into and across the Great Plains and to the Pacific Coast an entrepôt alignment grew up *not* on the Mississippi River, where in the beginning the places were basic retail trading centers more than wholesaling towns, but rather on the Missouri River and its logical structural continuations in the northern and southern Indian frontiers of the 1830s—Minneapolis–St. Paul, Sioux Falls, Sioux City, Omaha, St. Joseph, Kansas City, and Dallas–Fort Worth. On the Pacific coast the first alignment of entrepôts was coastal and in response to trade connections *maintained largely by sea.* That alignment has persisted up to the present despite the fact that intercoastal shipping has consistently decreased in importance and Pacific trade has never been of great national importance until very recently. Because of this sea link in the beginning the logical site for entrepôts in the West was on the coast. The fact that such a location today necessitates bringing products across the market area only to redistribute them in final trade backward into that area grows out of the persistent and conservative characteristics of wholesale trade.

WHOLESALING IN THE REGION OF SIMPLE AREALLY BASED URBANIZATION. As economic development advances, certain of the attributes of the pioneering environment are changed sufficiently to transform the pattern of trading places. Even in a relatively dispersed state, commercial farming or ranching will introduce a higher level of disposable income than is typical of pioneering economies; so, along with the greater number of customers, this relative affluence brings more and larger trading towns into being. The wide spread of settlement, however, tends to restrict the development of local trading centers at any close spacing. In much of the American West the cities are the only source, even in retail trade, for almost all lines of trade except food, gasoline, liquor, and other day-to-day demands. Such a concentration into moderate sized cities means that there is a relatively high correspondence in location between wholesale trade, of the sort that continues to move forward until it finally stops in quite advanced unraveling points, and most retailing. Because there is little support for cities other than trade, essentially all towns of any size will show up importantly in retail trade and the sort of wholesaling that needs only limited markets.

The very distance between these basic towns means that in any hierarchical arrangement the next higher level of central place would be far distant from most customers. That would encourage the continuance of the pioneering pattern of mail-order retailing or special ordering by retail merchants who have established business dealings with wholesalers located in a distant city. In this context the trading patterns of the Canadian Prairies serve us well as an example of how a widely dispersed population with a relatively high standard of living is supplied with a diversity of goods.

TRADING IN THE CANADIAN PRAIRIES. To understand the geographical situation of the Canadian Prairies is to conceive of the problems faced until railroads came; it is also to see why a particular wholesaling pattern grew up. The truth is that the Great Plains of the United States adjoin without interruption the broad level interior lowlands that we call the American Middle West. Thus, settlement advanced into the corn belt and continued, first across the Mississippi and then the Missouri. The string of towns that formed the entrepôt alignments followed.

In Canada the original settlement in Upper Canada was ringed to the north and west by the hard and watery Canadian Shield that made travel by water easy but settlement expansion difficult. The Great Lakes were a highway to their head, but the five hundred miles of Shield between Fort William and the Red River were difficult to cross with a valuable product, such as beaver pelts, and impossible to cross with agricultural produce. In that way the Prairies were effectively isolated and the intervening shield area all the way to Sault Ste. Marie was crossed but unsettled.

When the United States rail system reached to the Red River in 1871, there was a way out for the Canadian Prairies, and agriculture began to increase, stimulating the completion of the line of the Canadian Pacific Railway from Winnipeg to the Lakehead at Fort William in 1881. In either case the usable Prairies were first reached in the Red River valley in the last one hundred miles before it empties into Lake Winnipeg. When the Canadian Pacific settled on Winnipeg as its crossing of the Red River, the entrepôt of the Canadian West was fixed. In the short stretch of the Red River valley in Canada no alignment of entrepôts was feasible, so the Canadian Northern (built to the Lakehead in 1902) and the even later National Transcontinental Railway (through a branch opened to the Lakehead in 1908–11) used Winnipeg as their gateway to the Prairies and enlarged its entrepôt function. It is easy to understand how Winnipeg became the first emporium of the Canadian West and why it rose to the position of a true archetype of the unraveling point. It is not quite so obvious why it continues to hold that position nearly eighty years later, when the settlement fabric has expanded greatly and disposable income has grown on an even larger scale.

The discontinuity of the settled land in Canada combined with very late development in the Prairies to encourage a few large concerns to set out to apply the unraveling point idea to retail trade. The Hudson's Bay Company had always served as a "distributor" in this region, which had, until 1871, been their own Rupert's Land. T. Eaton and Company came west to join them. In either case, the idea was that of setting up a number of small shops or simple order offices to which a customer living in the dispersed farmlands of the West could go, choose his goods from a catalog, and order the goods for subsequent delivery. The orders were then collected into an unraveling point, usually Winnipeg. Indirectly this system encouraged the growth of Winnipeg as the predominant wholesaling city in the Canadian West by fixing there the unraveling point for a large amount of trade at the eastern margin of the area served.

The regional wholesaling pattern conforms to this general structure of trade. Except for the special needs of the region, in agricultural machinery for example, wholesaling is more concentrated in the Canadian Prairies than in the American corn belt where denser settlement and alternative supports for settlement and urbanization exist.

WHOLESALING IN THE AMERICAN GREAT PLAINS. The American part of the Great Plains is similar to the Canadian Prairies in its wholesaling structure, if less allied in retailing. A few relatively large towns do most of the wholesaling for the region. In this way the entrepôt alignment that began to develop on the Missouri River in the 1870s and 1880s has never been replaced. The sparse settlement on the Plains never engendered sufficient localized demand to make the forward projection of the alignment a practical undertaking. A few towns interior to the Great Plains took over wholesaling of staple commodities—so we find that Great Falls, Billings, Bismarck, and Rapid City sell these staples. Only in the south are there more important wholesale towns within the plains—Wichita, Oklahoma City, and regionally if not physiographically, Dallas and Fort Worth. The explanation is found in the greater amount of nonagricultural income and support in the oil producing area.

WHOLESALING IN A REGION OF RELATIVELY DENSE AGRICULTURAL SETTLEMENT WITHOUT INDUSTRY. When density of settlement in rural areas increases, the potential for urban growth is enlarged, particularly in the sense of allowing a greater number of cities within a given region. There will be more wholesaling centers, and their specific location will rest largely on transportation geography. The basic wholesaling towns will begin to show some resemblance to the net of towns envisaged under the operation of a central-place mechanism. But the more complex wholesaling town or towns will probably continue to reflect the pioneer pattern of the true entrepôt location. Thus, the port alignment, the river alignment, and the regional boundary site will tend to be perpetuated even after the forces that made the site distinctive may have ceased to operate.

The American South provides us with a clear example of this regional structure. There the rural settlement is quite dense, and there are many fundamental trading towns that grew up in the period of active nineteenth-century commercial agriculture. But the social and economic class structure was such that most of the considerable population could hope only to purchase the most staple of commodities. The small middle and upper class could not support a very elaborate trading structure, and there was little or no nonagricultural support to aid in the creation of a trading hierarchy. The result was the spreading of a rather even net of lowest order trading towns across the agricultural parts of the South. Above this basic net there was only a small number of wholesaling entrepôts at points where goods could be brought in from outside manufacturing regions and effectively distributed throughout the agricultural area. Since the entrepôt towns were either the coastal points from which early railroads were built inland—Charleston, Wilmington, N.C., or Mobile—or towns that had grown up around railroad junctions—Atlanta, Roanoke, Va., or Charlotte—most of the distribution was by rail; even today the major distributing points in the South are distinguished not by their central-place location but by their railroad importance. The small number of prime wholesaling towns that serve a large and fairly densely populated rural area have only recently been joined by settlements of different origin. In the Carolina Piedmont, industry now supports a large urban population.[5]

[5] The creation of a new extension of the manufacturing belt along the Interstate Highway that traverses the Piedmont of North and South Carolina (Interstate Route 85) is being bruited (see *New York Times*, July 9, 1966), and we may already see there the juxtaposition of important wholesale cities and factory towns with considerable numbers of people but no wholesaling.

WHOLESALING IN THE COMPLEX REGION. Economic complexity is present in all regions, even the most pioneering, but for simplicity and fundamental accuracy we have discussed the wholesaling structure of regions as progressing in complexity from the simple entrepôts typical of the geographically scattered and adventitious settlement of the pioneer stage to a highly complex wholesaling structure in diversified regions such as the corn or manufacturing belts. At the head of any order will stand the colonial (in the American context) entrepôt cities. These will have the greatest variety of wholesaling and will house the most specialized of trades wherein the forward movement of the unraveling point has never been practicable. In addition, these early entrepôts will tend to cling to those types of wholesaling in which the trading is in titles and other abstractions rather than in outright commodities. In the past the product itself may have been the item of trade, but so long as the abstract representation of that commodity suffices to establish and maintain a market, the city can keep the trade even when the products move in different lines of supply. In this way Boston can continue to trade in leather and wool, New York in cotton, sugar, and metals, and Chicago and Winnipeg in grains.

The leading cities of the complex economic region are also likely to be the places where taste is set for a country. That status gives them still another type of wholesale trade, that which depends upon the appraisal of style, as distinct from quantity. There is little judgment of style in the trading of wheat, but a great amount in the buying and selling of clothing in wholesale quantities. If style is set in the large, if not the single primate city, then wholesaling of style goods will tend to concentrate therein.

The geographical organization of wholesale trade begins to show itself clearly in such a complex region as the manufacturing belt. Cities there may be old entrepôts, such as Cincinnati or Boston, nineteenth-century manufacturing cities forming workplace towns such as Lowell or Akron, or even traditional trading centers, such as Springfield, Mass., or Lancaster. The settlement age and complexity in these regions also contribute another potential site for wholesaling in the major suburbs or satellites of the varied metropolitan cities. Thus, there are normally and generally at least four distinct types of city that could offer a site for unraveling.

The hierarchy of wholesaling in a complex economic region is real, but it is no mechanical derivative of central-place theory; rather, it is the result of the ontogeny of trading towns. *For that reason we must understand the history of wholesale trade to understand its present location.*

The Structure of Wholesaling: Progressive Internalization

It is not constructive to destroy the notion of central place as an analytical framework for understanding wholesaling without seeking to find a new organizing principle. For that principle we need ask, What is the element that gives order in the evidence cited? The answer becomes

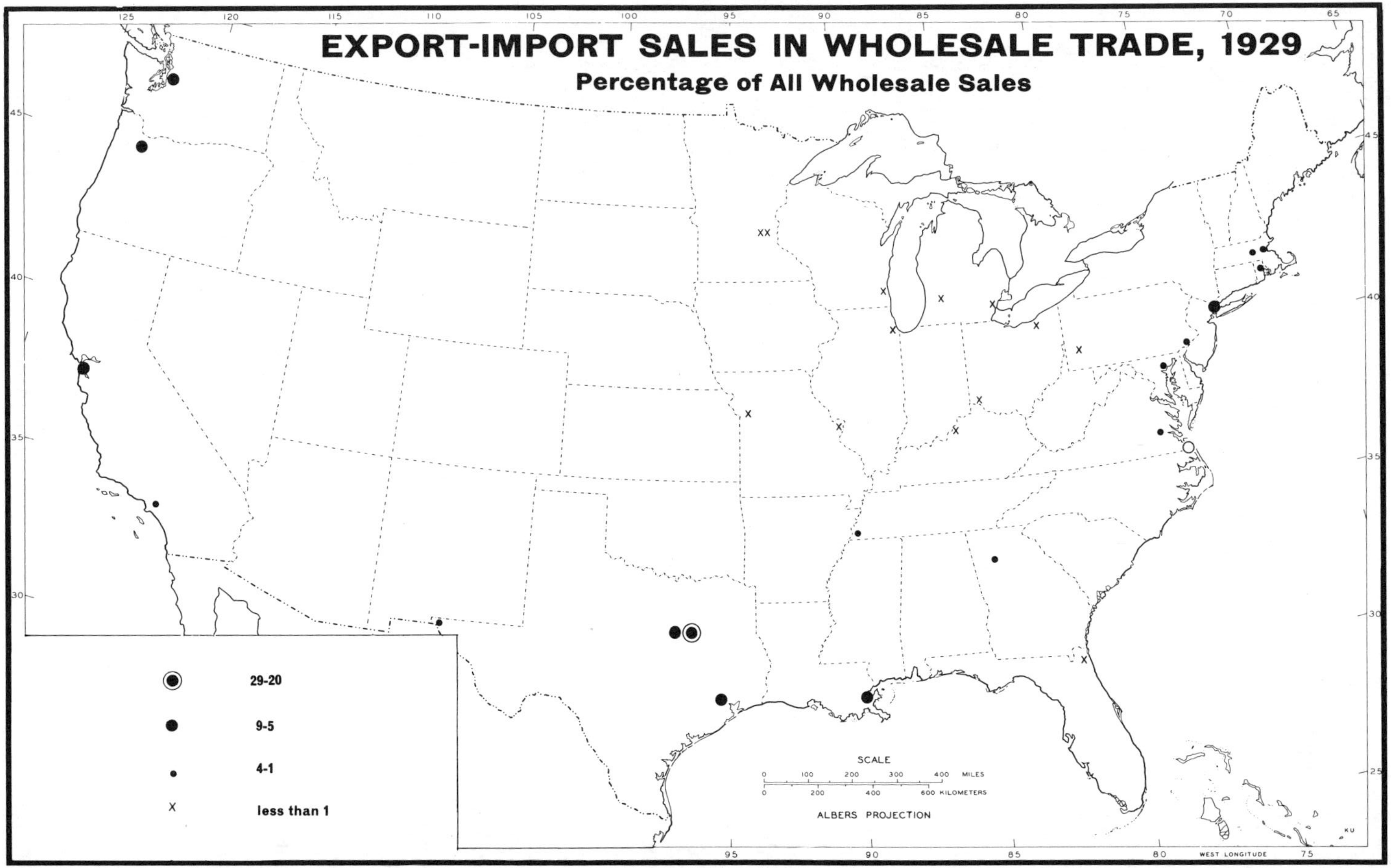
EXPORT-IMPORT SALES IN WHOLESALE TRADE, 1929
Percentage of All Wholesale Sales
29-20
9-5
4-1
less than 1
SCALE
0 100 200 300 400 MILES
0 200 400 600 KILOMETERS
ALBERS PROJECTION
WEST LONGITUDE
KU

Figure 11

clear, that there is a single dominant force at work that explains most of the contrast in characteristic wholesale locations throughout economic history in North America. That force can be called *the striving for internalization of wholesale trade.* The notion would hold that in the beginning trade was predicated on external conditions and lines of flow. The whole mercantilist development reviewed in Chapter 3 established this exterior causation as the base for trading development in the American colonies.

Obviously as any trading colony matures, it will tend to develop a commercial structure of its own, even if only to the extent that inland towns will come into existence to expand the realm in which external trade can be carried on. This is apparent when we look at South America or Australia today. There we encounter the second of the trading stages, a mature export economy with its continued concentration on entrepôts. To these dominant cities a moderate number of quite modest inland towns are added to broaden the region producing for export. But there is not the elaborated mercantile structure that we associate with even such avowedly rural areas as the American South. The small growth of inland towns does, nevertheless, represent the first symptom of internalization of trade. The towns are located at least in part in relation to domestic conditions, although still standing as the advanced depots of trade with the world at large.

Given potential for domestic development to a high level of economic complexity, a country will begin to carry on wholesaling that is not fully explained by external ties. Railroads will be built, inland junctions for those lines will afford good conditions for internal distribution, and these towns may have little or no function to perform that relates directly to importing or exporting. These trading junctions will become the first internally determined distributing centers—the geography of domestic trade may well become fixed and not subject to great change. Given a rural basic economy, the system is adequate to supply manufactures to a dispersed population and to collect the produce of that economy for transfer to other areas. Many countries in the world have not progressed beyond this rather fundamental geographical trading structure. The structure does not imply underdevelopment, but simple development, much akin to the Middle West in the middle of the last century.

The introduction of large-scale factory industry did not so much change this fundamental pattern as magnify it. The emphasis on internally derived trading locations was increased and the number of large population concentrations created by industry furnished possible locations for staple wholesaling.

The forces leading to an internally rather than externally derived structure for wholesaling supplement but do not supersede. Over the years, in both absolute values and relative measures, the original entrepôts have tended to hold their own as wholesaling centers. This is true even when they have lost much if not all their original maritime import and export functions—as it strongly the case in Boston and true to a lesser degree in San Francisco. The location of a city may have been related first to a particular configuration of trade; but as trade becomes internalized, the support of the wholesaling city becomes internalized. Still the city stays where it was first put by external forces. The very possibility of functional adaptation gives us a pretty strong clue to the important considerations in a theory of wholesale location. Obviously the notions of *impulse to trade* and the *agency of trade* are more fundamental than those of the range and threshold of goods that give the foundation to central-place theory.

Just as internal forces grow in importance as a nation develops, so those internal forces become progressively more significant as regions become developed. The first pattern will be dominated by the national transportation pattern; but as cities grow, the location of wholesaling among several cities will be related to intraregional considerations more than to extraregional forces. The first externally controlled wholesale cities will continue in use, but increments of trade will in the future tend to be parceled out among more cities and more in reflection of regional demands.

The Organizing Forces in Wholesaling

The organizing forces that affect the location, nature, and extent of wholesale trade are agency, impulse to trade, and the progressive internalization of location:

1. In the earliest stages of economic development, a parity between production and consumption allows the producer to be his own seller and the consumer his own purchaser.
2. Once local self-sufficiency is limited, there is no longer the automatic adjustment of production and demand, either in location or in timing, so that there have to be *agents of trade* seeking to rationalize the flow of goods from producer to consumer.
3. The idea of the *impulse to trade* must be substituted for the forces of restraint contained in the notion of upper and lower range of a good. The impulse to trade is, through the use of agents, capable of accumulation and periodic satisfaction.
4. Periodic demand satisfaction, normally through the office of agents, permits wholesale establishments to be located in a variety of places. Over the years, both in national and regional spatial contexts, and in urban structures, there has been a normally *progressive internalization of wholesaling location,* seeking to assure that increments to the wholesaling structure will be ever more responsive to domestic, and even local, demands.
5. The degree of internalization of trade locations depends for the most part on the density and complexity of regional, and metropolitan, development. *Thus, the maturing of trade practices joins with increases in population densities and heightened levels of disposable income to make distribution dominant in location decisions.* This is simply because the wholesaler's ties with his suppliers are formalized and maintained by practices that often require no personal contact, whereas when the wholesaler's body of customers grows in size and is more densely developed, real advantages exist in carrying on wholesale selling near that enlarged and clustered group.

The key to this internalization remains customer density. For the merchant selling his product to a thin and very disseminate market, the notion of proximity is meaningless. To cleave to one customer he must abandon another. In such a situation the ties of convenience to suppliers of goods may be more rewarding to the wholesale merchant, and he will tend to resist internalization of location found in a shift toward customers.

6. The culmination of the process of internalization comes in the branches of wholesaling wherein the service component is the strongest. The wholesaler who sets out most directly to serve his customer, through serving to hold inventory or replacements or through making final minor packaging adjustments, gains business by being close to those buyers. The converse is equally true, that merchants for whom customer service is unimportant but access to suppliers of great value, will tend to resist internalization. Most style-affected goods experience this dominance by the supplier and are characterized by a low degree of internalization.
7. In all types of wholesaling, transportation plays a vital role. Thus, technological changes in transportation are reflected in transformed depot locations. The impact of change in transportation is found much more clearly in the siting of wholesale establishments within metropolitan areas than in their location in space. With increasing internalization of wholesaling, trade establishments come to be located in terms of intrametropolitan convenience, which encourages siting exterior to the original city core. Those types of wholesaling that remain in the traditional core tend either to service customers in that core, representing a form of special internalization, or else to engage in those types of wholesaling characterized by customer conflux on the merchant's establishment. This, again, when located in the central business district or its environs, represents a specialized application of the notion of internalization; we may account for most of the locational dynamism in wholesale selling through this single force.

1929: The First National Portrait of Wholesaling

All our foregoing efforts to sketch the national pattern of wholesaling have had to depend upon indirect and incomplete evidence. Such qualities, however, stand as restrictions rather than outright disqualifications in the use of the data. In 1930, data of greater completeness and competence were collected in the Fifteenth Decennial Census. These reflect conditions of business during calendar 1929, a year that provides one of the more exceptional bases for a historical series. Nonetheless, some virtue attaches to the year in the sense that it represented the fatal incandescence of the pure laissez-faire economy. In measures of sales, business establishments, and employment, 1929 recorded conditions not to be exceeded until the business census of 1948. But in the picture of trade location there was a continuous and only slowly transforming pattern. Thus, no particular problem is raised when we must base our first portrait of American wholesaling on the year of the Crash.

The most immediate impression conveyed by that portrait is one wherein the colonial forebears are still suggested. In 1929 the first ten cities in wholesale sales were the archetypal entrepôts: Boston, New York, and Philadelphia on the East Coast; Detroit and Chicago on the

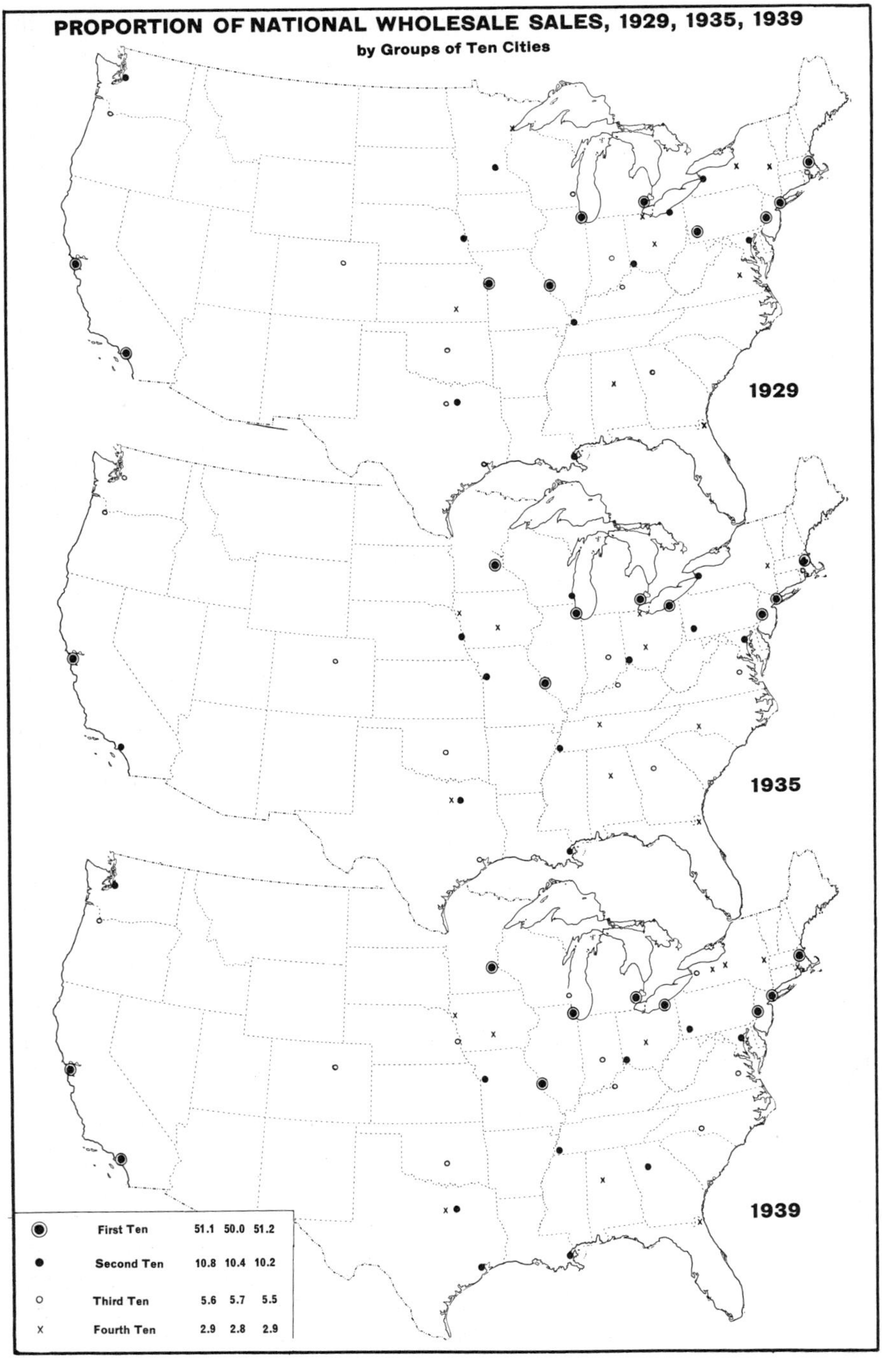
PROPORTION OF NATIONAL WHOLESALE SALES, 1929, 1935, 1939
by Groups of Ten Cities
1929
1935
1939
First Ten 51.1 50.0 51.2
Second Ten 10.8 10.4 10.2
Third Ten 5.6 5.7 5.5
Fourth Ten 2.9 2.8 2.9

Figure 12

Lakes route; Pittsburgh, St. Louis, and Kansas City on the River route; and the two West Coast gateways, San Francisco and Los Angeles. In the absence of metropolitan statistical districts, the data used are approximations combining the figures for cities that have subsequently been joined in hyphenated districts (e.g., San Francisco and Oakland) but no attempt was made to include all places now in metropolitan districts. Only in New York was the test of hyphenation modified to include New York itself, Jersey City, Newark, and Paterson. In these top ten wholesaling cities of 1929 half the trade was carried on (51.09 per cent), and each city accounted for 2 per cent or more of the total U.S. trade. To understand such domination of the many by the few we must reflect briefly on the difference between wholesale and retail trade.

WHOLESALE PURCHASES AND THE ELEMENT OF CHOICE. The key to this distinction is found in the matter of customer gravitation. In selling to retail customers there is the assumption that those people are making choices that require their shopping, a term that suggests the locale in which it is carried on. But in wholesaling only in certain types of transactions are choices made. *The wholesaling transaction, in essential form, is probably as much a decision with respect to quantity as it is of type of good.* Under those circumstances, the retail merchant often need not view the goods as much as he must assess the temper of his customers. That judgment can be formed better behind *his* counter than in front of his wholesaler's. The result is that wholesale trade can be concentrated to a degree that retailing cannot. The location of such concentration is a separate question, the answer to which is particularly one of time precedence. The largest ten cities in 1929 were the ones that rose first to prominent trading positions. This may be said another way; certain regions became fruitful for trade earlier than others. New England, the Middle Atlantic States, the Middle West, and the urban sections of California assumed importance before the South, the Mountain West, or the Northwest. As those other regions rose in importance they faced two obvious facts: (1) for those types of trade in which concentration is economically desirable, the entrepôts were already established, and (2) for those types of trade in which service assumed a greater role, the newer enlarged local market could permit the creation of second-rank wholesale towns.

The Second Rank in 1929. The second rank of ten wholesaling cities was quite clearly comprised of places commanding neither the tributary area nor the over-all trade of the first ten cities. The second group accounted for only 10 per cent of the national wholesale trade and showed a strong tendency toward location on the margin between the fully developed manufacturing and corn belts and the emerging regions. Four of the cities served to complete the traditional alignments of the entrepôts in the Middle West: Baltimore and Cincinnati in the River alignment and Buffalo and Cleveland in the Lakes structure. This second rank also began to outline the north–south alignments of towns in which wholesaling was first brought to the interior: New Orleans,

Memphis, and Minneapolis on the Mississippi (as well as the original interior town, St. Louis); and Dallas and Omaha, which joined with Minneapolis and the original frontier town, Kansas City, to form a plains border alignment between the Gulf and the Canadian boundary. The final member of the group, Seattle, anchored the northern end of the West Coast entrepôt alignment.

The Other Ranks in 1929. The third rank of ten cities is made up of two components, those cities that were regional capitals in underdeveloped regions or those cities that were secondary wholesaling towns of some antiquity but obvious truncation of function. Portland, Oreg., Providence, Milwaukee, and Louisville all were rather old "ports" without really large hinterlands, whereas in 1929 Houston, Atlanta, Denver, and Oklahoma City dominated growing but not fully grown areas.

The fourth rank was made up of minor members of the several alignments—Norfolk and Richmond in the string of older East Coast entrepôts; Albany, Syracuse, Toledo, and Duluth–Superior in the Lakes structure; Wichita in the Plains Border string; and Birmingham in the string of railroad-based towns on the inner edge of the coastal plain. Columbus, Ohio, and Jacksonville each stood somewhat apart as the secondary town of a region or the entrepôt for a developing land.

AND AGAIN IN 1939. Structurally, this last prewar business census begins to suggest impending developments. The South was becoming more economically developed, shedding some of the pioneering cast that had formerly restricted wholesale trade to a couple of largely exterior trading cities. The strongly rail-oriented unraveling points on the Plains Border began to show some signs of decline, a decay that suggested the ultimate growth of substitute unraveling points in the Mountain West. But in 1939 the decline more than the replacement was in evidence. Still, in the Far West the pioneer pattern persisted, with the early ports dominating a region of growing productivity to the extent that two of the four cities stood in the first rank and another in the second. This was no nesting along hierarchical lines, for the evidence is clear that each city tied backward to the East relatively independent of its Pacific neighbors.

Wholesaling in Metropolitan Areas

The figures used for this simple comparison of American trading cities had to rely on information for political cities through 1939. The first time data by metropolitan geographical units were published was in the Census of Business, 1948, which contained information on trade in the Standard Metropolitan Areas that had been defined in preparation for the 1950 population census. In 1948 the ten top wholesaling metropolises remained the same as before and continued to account for half the nation's sales. In the second rank of ten cities there was some shifting, due certainly in part to the use of full metropolitan areas rather than merely central cities. After dropping out in 1939, Omaha bounced

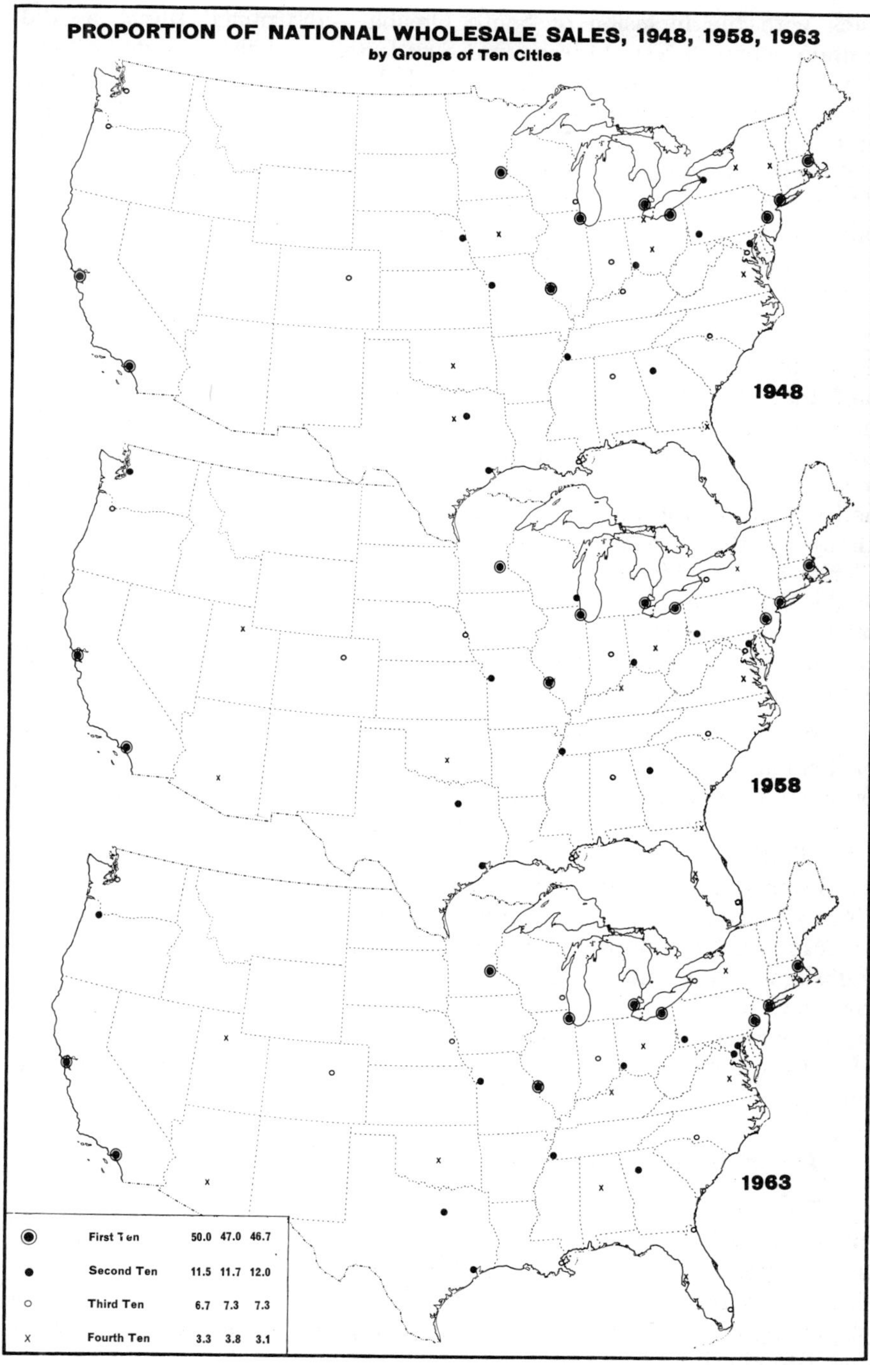
PROPORTION OF NATIONAL WHOLESALE SALES, 1948, 1958, 1963
by Groups of Ten Cities
1948
1958
1963
First Ten 50.0 47.0 46.7
Second Ten 11.5 11.7 12.0
Third Ten 6.7 7.3 7.3
Fourth Ten 3.3 3.8 3.1

Figure 13

back with the inclusion of South Omaha and Council Bluffs, as did Buffalo, whereas New Orleans and Seattle reflected the downtown location of their port concentration of wholesaling by bulking less large when the whole metropolis was tallied. In the third and fourth groups of ten there was modest change. Washington joined the ranks of the forty top wholesaling centers, while Sioux City departed, and the first thirty cities continued to account for two-thirds of all American wholesale trade. The structure remained essentially identical to that of 1939, with a strong alignment of cities in accordance with the historical geography of transportation.

A LOOK AT 1958. The 1954 census of business did not furnish a figure for the total wholesale trade of metropolitan areas, so we must omit that year and go on to 1958 when the Bureau returned to its former practices. In the decade 1948–58 the underdeveloped sections of the country began to disappear, if we may judge from wholesale trade location. The top ten cities remained the same, but their contribution had begun to drop to less than one-half (47 per cent). There was little change in the second rank of ten, but in the third and fourth groups Florida and the Mountain West had begun to assume greater importance while the minor trading centers of the manufacturing belt had lost relative status. Clearly the United States was moving to a more direct equation of space and selling, which we might consider more in tune with notions of trading centers.

AND FINALLY 1963. Most of the postwar transformation of the group of dominating wholesale cities had taken place by 1958. The most recent business census to be published, 1963, shows the persistence of the earlier pattern with only such slight readjustments as the rise of Jacksonville, Portland, Oreg., and Washington to take places formerly occupied by Birmingham, Seattle, and Milwaukee. There were no changes in the cities involved merely these shifts of relative rank. The percentage contributions of the groups are also shifted slightly, with the top ten furnishing slightly less and the second and third groups slightly more. The largest thirty cities still accounted for just two-thirds of the national total. In this way we can see the continued but slow ebbing of the dominance of the traditional entrepôts and the rise of the regionally based trading centers.

The Pattern of Employment

There is no doubt that any single measure of activity in a distributing industry has its faults, but it seems that employment in wholesale trade is perhaps the most standard measure. Wages are subject to considerable regional variation even if jobs are identical, but number employed is only slightly skewed by low wage rates and gives an easily understood criterion of comparison.

WHOLESALE EMPLOYMENT IN 1929. No surprise should attach to the statement that there is great similarity in pattern between wholesaling

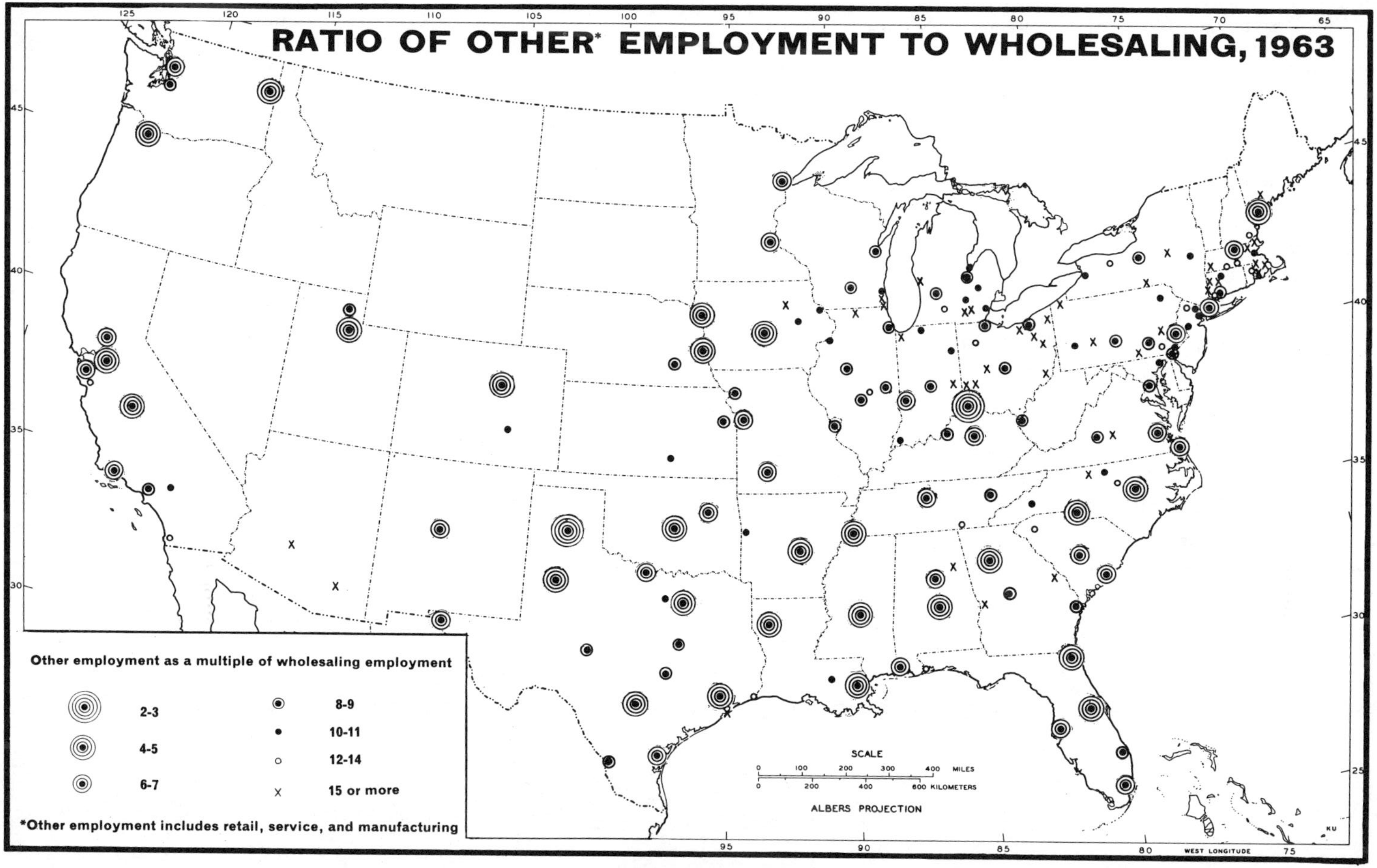

RATIO OF OTHER* EMPLOYMENT TO WHOLESALING, 1963
Other employment as a multiple of wholesaling employment
2-3
4-5
6-7
8-9
10-11
12-14
15 or more
*Other employment includes retail, service, and manufacturing
SCALE
0 100 200 300 400 MILES
0 200 400 600 KILOMETERS
ALBERS PROJECTION
WEST LONGITUDE

Figure 14

Table 6. Proportion of Urban Employment in Wholesaling (For SMSAs) (1963)

Proportion of Urban-type Jobs Found in Wholesale Trade: $\dfrac{\textit{Retail + Service Trades + Manufacturing Employment}}{\textit{Wholesale Employment}}$

1 Job in 3 in Wholesaling		*1 in 8*		*1 in 13*	
Amarillo		Boston	Madison	Beaumont-Port Arthur	Lancaster
Cincinnati		Champaign-Urbana	Roanoke	Chattanooga	Lima
		Chicago	St. Louis	Decatur	Paterson-Pasaic
1 in 4		Grand Rapids	San Angelo	Durham	Providence
Charlotte		Green Bay	Savannah		San Diego
Fresno		Harrisburg	Tacoma		
Jacksonville		Indianapolis	Toledo	*1 in 14*	
Lubbock		Lincoln	Topeka	Bridgeport	Springfield, Mass.
Orlando		Laredo	Waco	Kalamazoo	Trenton
Raleigh			W. Palm Beach	Rochester	Wilkes-Barre
				San Jose	Worcester
1 in 5		*1 in 9*			
Atlanta	New Orleans	Altoona	Macon	*1 in 15*	
Dallas	Oklahoma City	Austin	New Haven	Binghamton	Phoenix
Denver	Omaha	Cleveland	Peoria	Brockton	Rockford
Des Moines	Portland, Me.	Columbus, O.	Saginaw	Jackson, Mich.	Waterloo
Houston	Portland, Oreg.	Fort Wayne	St. Joseph		
Jackson, Miss.	San Antonio	Huntington-Ashland	Springfield, Ill.	*1 in 16*	
Little Rock	Sioux City	Knoxville	Syracuse	Akron	Erie
Memphis	Sioux Falls	Los Angeles	Washington	Augusta	Lewiston-Auburn
Montgomery	Spokane	Louisville	Wilmington, Del.	Columbus, Ga.	Lynchburg
	Stockton			Dayton	Stamford

1 in 6
Albuquerque
Allentown
Columbia
Duluth-Superior
El Paso
Kansas City
Lexington, Ky.
Miami
Minneapolis-St. Paul
Nashville
New York-NE N.J.
San Francisco-Oakland
Santa Barbara
Springfield, Mo.
Tulsa
Wichita Falls

1 in 7
Birmingham
Charleston, S.C.
Corpus Christi
Fitchburg-Leominster
Mobile
Norfolk
Richmond, Va.
Sacramento
Seattle
Tampa-St. Petersburg
Terre Haute

1 in 10
Asheville
Bay City
Davenport-Rock Is.
Detroit
Dubuque
Evansville
Fort Smith
Fort Worth
Greensboro-High Point
Hartford
Lansing
Newark
Ogden
Philadelphia
San Bernardino-Riverside
Scranton

1 in 11
Albany-Schenectady
Baltimore
Baton Rouge
Buffalo
Cedar Rapids
Flint
Jersey City
Pittsburgh
Pueblo
South Bend
Wichita

1 in 12
Greenville
Manchester
Pensacola

1 in 17
Galveston
Johnstown, Pa.
Springfield, O.
York, Pa.

1 in 18
Canton
Muskegon
Newport News
Reading
Winston-Salem
Youngstown

1 in 19
New Bedford
Utica-Rome

1 in 21
Racine

1 in 23
Fall River

1 in 24
Lawrence-Haverhill
Muncie-
Pittsfield

1 in 26
Gadsden

1 in 29
Ann Arbor
Lowell
Waterbury

1 in 30 or more
Hamilton (30)
Tucson (31)
Gary-Hammond (32)
Kenosha (34)
Lorain-Elyria (37)
Steubenville (41)
New Britain (48)

employment and sales. The ten leading cities in employment are identical with those in sales, but the contribution differs. The first ten in sales account for 51 per cent of the national total but their stock of wholesaling jobs supplies no more than 43.5 per cent of the U.S. total. The disparity completely disappears in the second, third, and fourth ranks of ten cities, so it is only the largest cities that sales run well ahead of employment. To some degree the high level of selling comes from sales agencies and other facilities that carry on wholesaling transactions without labor-using warehousing. But also the economies of scale favor the large cities in the efficiency of the use of workers.

AND IN THE THIRTIES AND FORTIES. It is unnecessary to follow through each census in detail to note the significant features of wholesale employment. *In the time series available it is clear that employment tends to furnish a prediction of conditions in sales.* The contribution of the largest ten cities was in sales but the employment predicted the course of sales. In geographical terms, the level of employment shows a tendency to anticipate sales by about a decade. The rise of trading cities in Florida, Texas, and the Mountain West shows up in employment about a decade before it shows up in sales, suggesting that the conditions for successful wholesaling are foreseen and the trade is encouraged by merchants in emerging areas before the market is fully developed. This would support the idea that tradition plays an important role in these trade linkages and acts as a conservative force, keeping the sales in the more distant traditional entrepôt for perhaps a decade after the marginal volume of localized trade for internalization is reached.

WHOLESALE EMPLOYMENT IN 1963. If we look upon employment as a predictive device in wholesaling, the data for 1963 suggest little change. The dependence upon the traditional entrepôt remains, and little development in California or the West in general is forecast. This does not mean that the cities in these areas will fail to contribute to the structure of selling—they already do so in the lower ranks of wholesaling—but it does mean that the primacy of a very small number of really dominant cities will persist.

WHOLESALING EMPLOYMENT IN RELATION TO TOTAL EMPLOYMENT. What are the characteristics of wholesaling employment in the context of all employment? To answer that, data on Standard Metropolitan Statistical Area (SMSA) employment in manufacturing, retail trade, and service trades have been compared with employment in wholesaling. In Table 6 the results are shown, with the cities classified in terms of the number of times larger the manufacturing–retail service employment is than the wholesale employment. The cities relatively weakest in wholesaling show up as those with the highest level of multiplication. ...benville, Hamilton–Middleton, Lorain–Elyria, New Britain, Gary, ...owell, Gadsden, and Tucson have more than 25 times as ... employment as that in wholesale trade. Save for Tucson, these cities are obviously industrial satellites with little attraction for the surrounding rural population, however desperate it may be for trade.

In general terms the relative importance of wholesaling employment reflects not any assumed hierarchical position of the city, but rather its dependence upon trading for support. Outside the manufacturing belt trade takes on greater importance by the absence of alternative forms of employment. We find in the emerging manufacturing belt of the Carolina Piedmont that even such a specialized furniture wholesaling SMSA as Greensboro–High Point does not stand notably high in general wholesaling because there is large manufacturing employment to expand the basis of comparison, but Raleigh and Charlotte, each with less manufacturing than Greensboro–High Point, class as wholesaling centers.

The cities most dominated by wholesaling employment in relative terms are the great regional centers in agricultural areas. The highest include southern trade centers such as Charlotte, Raleigh, Jacksonville, and Shreveport, and those on the border of the South–Lubbock, Amarillo, and Cincinnati. Outside this group of southern cities only Salt Lake City has 25–35 per cent of the workers in wholesaling employment. With between 16 and 20 per cent in wholesale employment come the southern and western trade centers of first stature such as Montgomery, Miami, Omaha, Dallas, Denver, Portland, Oreg., San Francisco–Oakland. Portland, Maine, shows functional affinity for this nonecumenical group, while New York City is elevated to the second rank of wholesaling towns by its massive apparel wholesaling structure. Only when we turn to an employment proportion in wholesaling of one in seven or eight do even a few of the trading centers of the manufacturing belt show up. Boston, Harrisburg, Toledo, Indianapolis, Grand Rapids, Chicago, and St. Louis show up as much as trading centers as they rank as manufacturing towns, and along with the lesser wholesaling centers of the West and South (Seattle, Tacoma, Tampa, Savannah, and Charleston, for example), they make up the third group with 10–12 per cent of employment in wholesaling.

Those cities with 10 per cent or less employment in wholesaling lie mostly in the manufacturing belt and represent the cities that are normally cited as comprising hierarchies of central places. They may fulfill that function in retail trade, but in wholesale trade they show no specifically hierarchical structure.

The Components of Wholesale Trade: 1958

The generalized pattern of wholesaling presented in the preceding chapter tells us location of business activity but not of specific merchants. No doubt in comparing cities we gain the best measure of wholesaling through the use of such aggregated information. When we wish to know the background of the spatial pattern, however, we need to know the components that join to provide the totals. This is particularly the case when we reflect that in wholesale trade there may be no functional convenience derived through purchase of all goods in the same city.

The use of aggregate data for retail trade has considerable justification since retail purchasing is normally carried on by the customer through the physical act of shopping. Because he cannot be divided, neither can his trade, at the same moment at least. In wholesaling there is normally an agent of trade whose operations permit the customer to divide his custom quite handily.

In wholesale trade we cannot make the automatic assumption that customers converge on the trading establishment. Quite clearly in many lines the customers seldom, if ever, visit the merchant's place of business. Thus, wholesaling is at heart definitely, and dominantly, the distribution function and lives not by the rules of trade in general, but by the rules of delivery. On the other hand, certain types of wholesaling require the attendance of customers at the merchant's place of business, and these are greatly concerned with the efficiency of customer convergence. Anticipating our story a bit, it should be evident that the pure distributor can function in the middle of a corn field whereas the wholesaler with customer convergence cannot.

CONVERGENCE AS A CONSIDERATION IN WHOLESALE LOCATION. The basis for the convergence of customers on a single wholesaling center is found in the exactions of time. In all types of wholesaling the ability to anticipate the *quantities* that will be in demand and the *specific types* of goods that will be bought is important. In some types of wholesaling the latter concern is not a simple one of apples and oranges but rather one of what color apples and what shape. To carry on a business in which style plays a part, a retailer must not only judge the future tastes of his customers but also must know what offerings of goods manufacturers will make to satisfy that taste. If the retailer seeks to forecast what his customers will want during the next season, he combines a geographically specific knowledge—of their taste—with a time specific knowledge—of what the manufacturers will offer during that season. To gain the latter knowledge he must visit an unraveling point of sufficient size to permit the *appraisal of the offering.*

The appraisal of the offering varies with the type of goods and the class. Taking clothing as an example, we may note that clothing is probably the most varied of commodities. Some types of clothing, particularly men's underwear and handkerchiefs, are subject to only the most modest secular change in taste. In this case the strong development of brand labeling has obviated the necessity for the specific appraisal of the offering. No retailer travels to Chicago to buy men's T-shirts even though he might buy them adjunctively while there, if offered particular price incentives. On the other hand, in what are called fashion clothes it is critical for the retailer to judge the offering before making his purchases. Some of this judgment may be exercised through the agency of salesmen, as it is by small retailers unable to be away on shopping trips; but any store seeking to present a strong image of timeliness is unlikely to depend on such a system with its susceptibility to manipulation. For those traders the buying trip to the *center of wholesale convergence* is a necessity.

Unfortunately the strictures of disclosure in the census of wholesale trade make it difficult to compare SMSA's in detailed lines of goods. We must use instead the general class, dry goods and apparel in this case, to serve as indicator. Even then we encounter the sort of niggling rigidity that mars the census, in that data are withheld for Chicago despite the presence there of 621 establishments in the trade in 1958. Within these limits we still are able to gain a general impression of the location of major convergence in the wholesale sale of apparel.

In dry goods and apparel selling it is necessary to reduce the impact of local population contrasts so we may judge possible convergence. To do this, the data are shifted from an absolute to a per capita basis, as shown in Table 7. With such a base certain SMSAs stand out. Those with a per capita value for dry goods and apparel wholesaling of $100 or more are as shown in Table 7.

Table 7. 1958 Per Capita Wholesale Trade in Dry Goods and Apparel SMSAs

All with $40 Per Capita or More

New York	$829.5	Scranton	$63.7
Greensboro	225.3	Lynchburg	63.6
Brockton	222.5	Cleveland	62.5
Atlanta	200.4	Fall River	58.6
Manchester	174.4	Philadelphia	55.5
St. Louis	166.6	Knoxville	48.7
Dallas	161.3	Kansas City	47.4
Boston	150.0	Portland, Maine	46.8
Nashville	113.6	Chattanooga	45.4
Charlotte	90.6	Cincinnati	43.5
Los Angeles	89.7	Springfield, Mo.	43.1
Memphis	80.8	Allentown-Bethlehem	42.6
Lawrence-Haverhill	76.7	Indianapolis	42.4
Fort Smith	72.4	Miami	41.3
New Bedford	69.2	Minneapolis	40.7

The pattern shown is comprised of the truly dominating traditional wholesaling centers, where we might expect convergence to take place. On the basis of other evidence, it seems probable that Chicago would be in this group if we had the data, so we end up with New York and Boston on the East Coast, Chicago and St. Louis in the Middle West, Atlanta and Nashville in the South, and Dallas in the Southwest. Greensboro, Brockton, and Manchester are manufacturing cities for dry goods and apparel, so they show up not for convergence but for wholesaling operations intimately tied to production.

We may introduce the notion that there are three types of wholesaling cities in the clothing and dry goods trade. There is the single dominant style-setting center. In the United States this is New York but it has several rivals in the world as a whole, notably Paris for women's clothing and London for men's. Next in order of size of geographical tributary area come the major regional dry goods wholesaling centers where retailers come to look over goods on occasion and from which salesmen diverge to tap the territory. Finally, there are the manufacturing-tied wholesaling centers whose location is completely derivative. Boston, St. Louis, Dallas, Brockton, Haverhill, Chattanooga, and Fort Smith are members of this class, although Boston, Dallas, and St. Louis also are major regional dry goods centers. The combination of derivative and regional selling elevates the two above the level of other regional centers.

THE SUMMIT OF CONVERGENCE, THE SPECIAL CASE OF NEW YORK. The impact of style is basically non-hierarchical and somewhat independent of area. It might be argued that there are regional styles in a country as large as the United States or Canada, and that assertion can be supported. Nonetheless, these areal variations are made obvious by their limited occurrence; in most situations the national taste dominates. To have such a national taste there must exist an *arbiter market* in which taste is established as price and supply are in other markets. There is no mechanistic rule that there can be but one arbiter market or that the decision is laid down for all goods in a single market.

In goods with a component of style, that style is fully as much a consideration in the "utility" of the garment as is the staple from which it is made. The Fall River clothier probably wants to see the New Bedford goods alongside those of Scranton and Manchester before he buys any of them. So New York dominates apparel selling because it stands as the arbiter market, not because it is a particularly efficient or easy place to sell. If that were the test, there would be apartments lining West 34th Street. Instead of the homeward journey of workers, however, there is a massive convergence of buyers from all over the United States.

To understand convergence we may turn to a feature of the *New York Times* that is literally buried in the back pages. There on most weekdays are listed "Reports of Arrival of Out-of-Town Buyers in New York," detailing by city of origin and types of goods to be purchased those representatives of retailers, and a few wholesalers, coming into New York on the particular day. This list is certainly not complete. It obviously tallies only those buyers furnishing the locale of their stay in New York and the nature of their buying to the *Times*. This means that the preponderance of reports falls in the apparel categories, although other buyers appear to seek listing particularly when there is the convergence of a specific category of buyers, such as gift shop agents or toy purchasers normally during the tenure of a trade show. Also there is an obvious underappraisal of merchants coming from the immediate vicinity who have easily established connections.

The period of study covers the eight months beginning with November 1965 and concluding with June 1966. During that period the report for one day during each week was analyzed. The result is that in a period when some 37,554 buyers were tallied by the *Times* the sample analyzed amounted to just over ten thousand (10,033), a sample of 26.7 per cent of all buyers listed in the *Times*.

The sample group demonstrates anew the truth that the density of trading is greatest in the manufacturing and corn belts. Six states—New York, Pennsylvania, Ohio, Illinois, Michigan, and Massachusetts—provide half the buyers visiting New York, but these states have only about one-third the U.S. population. The big states with the large cities are dominant partly because it is particularly the large city department stores that send buyers to New York. On an average day Philadelphia, Pittsburgh, Cleveland, Detroit, Boston, and St. Louis will have department store buyers in New York. Interestingly, Chicago is not normally represented by department store buyers.[6] Some of the medium-size American cities will be represented, but the tendency is less in that group than in the really big cities. In fact, the smaller but rather enterprising cities—Raleigh or Baton Rouge or Peoria—seem more likely to send buyers to the arbiter market than do the medium-size industrial cities of the Northeast.

The pattern is suggestive of ideas advanced by William J. Reilly nearly forty years ago. In trying to explain the retail gravitation in Texas [7] he noted, along with his much more widely known view that customer flow and gravity live by the same physical rules, that there is a tendency for the more well-to-do in a city to travel to the next higher order of place to do their shopping, while the majority of people go to the nearby downtown itself. In the use of the style-creating mechanism of New York, a few cities seem strangely underrepresented. Chicago is the prime example, but Dallas, Los Angeles, San Francisco, Montréal, and Toronto might be noted. To some degree distance may account for the California cities, but Chicago and Dallas are no farther away than cities that are well represented. There is in this the suggestion that a few cities try to be their own arbiters of style, and those cities are the ones most cosmopolitan in character and least likely freely to grant dominance to New York. This is probably the explanation of the minor interest of Canada in New York at the same time that much of the Canadian market is saturated by American products. In Canada there is a tendency to pass over New York and look directly to Europe when it comes to style and taste rather than practical operation.

When we adjust the flow of buyers to New York to the population size of states we find that New England, the Middle Atlantic states, Ohio, and Missouri still show up strongly in the flow of buyers. Florida, Louisiana, Texas, and all the West become minor contributors. Despite its vast population, California stands eighth from the bottom, bracketed by a number of very rural states not noted for their obsession with fashion. Illinois is only slightly less divorced from the New York market.

6 The Editor of this series, whose intimacy with the mind of Chicago is unmatched, suggests that its citizens consider that Chicago's size and self-confidence make the city its own arbiter.

7 William J. Reilly, *Methods for the Study of Retail Relationships, University of Texas Bulletin*, Austin, 1929. Reprinted as *University of Texas Studies in Marketing*, ss Research, 1959.

Thus there is convergence on New York from all over the country, and it is some measure of this movement to note that in 1964 the *New York Times* published the arrival of fifty-six thousand buyers and in 1965 fifty-one thousand.

OTHER ARBITER MARKETS. This is not the place to attempt a full accounting of the arbiter markets of the United States or of the world as a whole. It is an appropriate place, however, to mention some of the more significant examples to demonstrate that the case of New York is not unique. Furniture shares with apparel a great influence of style on sales, but the timing is different. It is necessary for the retailer to know the trend of the market and the items introduced each season. Unlike clothing, however, the turnover rate is much slower, measured in years rather than months, so even though a merchant may be aware of changes and of introductions he need not respond to them immediately. In this way the regional furniture market can serve to create reasonable approximations of the national taste, and the producer-tied wholesaling center can compete with the major metropolitan centers. The fact that housing varies from one part of the country to another has also contributed to the parceling out of the furniture wholesaling. San Francisco and Los Angeles thus can have their own furniture marts, as do Chicago and New York. Grand Rapids and High Point, N.C., have elaborate furniture wholesaling organizations based not on potential nearby customers and traditional wholesale city status, but on a large volume of local manufacturing. All these furniture markets, however, depend upon the convergence of customers and the creation of a regionally based arbitration of taste.

To delineate the nature of such a regional arbiter market we may look briefly at furniture wholesaling in San Francisco which concentrates in the Western Merchandise Mart. There visitors must register so it is possible to analyze their origin. In a recent tally the breakdown shown in Table 8 was found. The pattern shown is the sort that we would anticipate finding ranged around a major regional arbiter. Nearly two-thirds of the buyers converge on San Francisco from northern California itself but southern California, Oregon, Washington, Idaho, and Nevada also spend appreciable numbers.

In both these cases the arbiter city is a true entrepôt, and the earliest to be established in its particular regional context. This is not surprising, for we would anticipate that *primacy of time is a factor in the creation of primacy of taste,* and the facilities for convergence are normally best when focused on the initial entrepôt.

This discussion of the arbiter market serves to explain another force at work on the structure of American wholesaling and a force that, like the historical linkages, tends to interrupt the mechanical shaping of the pattern that would come with abstract central-place dynamics. The impact of the time represented in the entrepôt is joined with the impact of arbitration of taste to create a distinct geography of wholesaling quite independent in origin from the typical retail pattern.

Table 8. Market Area Breakdown of Buyer Registrations, Winter Market (1961)

Geographical Area of Origin of Buyer at Western Merchandise Mart	*Number of Buyers* *	*Percentage of Total of Buyers*
San Francisco	2,292	10.32
Oakland	1,463	6.59
Other Northern California	10,119	45.59
(Total Northern California)		(62.50)
Los Angeles	656	2.95
Other Southern California	936	4.33
(Total California)		(69.78)
Oregon	1,830	8.24
Washington	1,738	7.83
Idaho	992	4.46
Utah	698	3.14
Nevada	396	1.78
Montana	228	1.02
Arizona	194	0.87
Alaska	79	0.35
Colorado	73	0.32
Canada (probably British Columbia)	70	0.31
Texas	68	0.30
Hawaii	65	0.29
New York	39	0.17
Illinois	29	0.13
Massachusetts	19	0.08
Ohio	15	0.06
Missouri	13	0.05
Michigan	8	0.03
Pennsylvania	8	0.03

Also: Arkansas, Connecticut, Delaware, District of Columbia, Florida, Georgia, Indiana, Iowa, Kansas, Kentucky, Maryland, Minnesota, Nebraska, North Carolina, North Dakota, Oklahoma, Tennessee, and Wisconsin. Totaling 70. Foreign 5.

Source: Western Merchandise Mart, San Francisco

* Total number of visitors to Winter Market, 1961 = 22,194.

Wholesaling in the National Frame

We have already discerned several of the forces that shape wholesaling location patterns. Are there others? To answer that question it is necessary to carry out a brief appraisal of relationships of population and wholesale trade. To accomplish this we may in simple fashion, plot certain characteristics of American wholesaling against population (Fig. 15). Thus, the *Y*-axis is used to plot the several characteristics to be considered, and the *X*-axis remains the measure of population in SMSAs in 1958.

EMPLOYMENT IN WHOLESALING AS A PERCENTAGE OF SMSAS POPULATION. On the face of it we would assume that there would be a constant

EMPLOYMENT IN WHOLESALING AS A PER CENT OF SMSA POPULATION, 1963

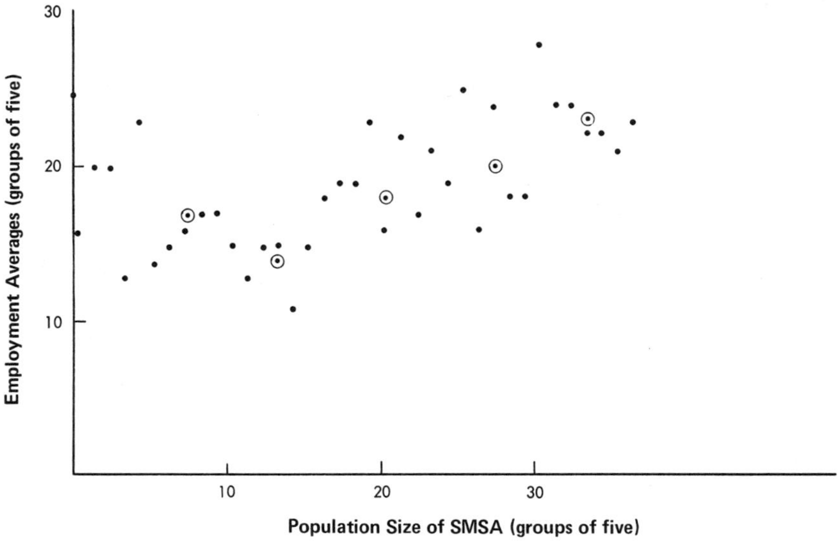

PER CAPITA WHOLESALE SALES RANKED BY POPULATION SIZE OF SMSA, 1963

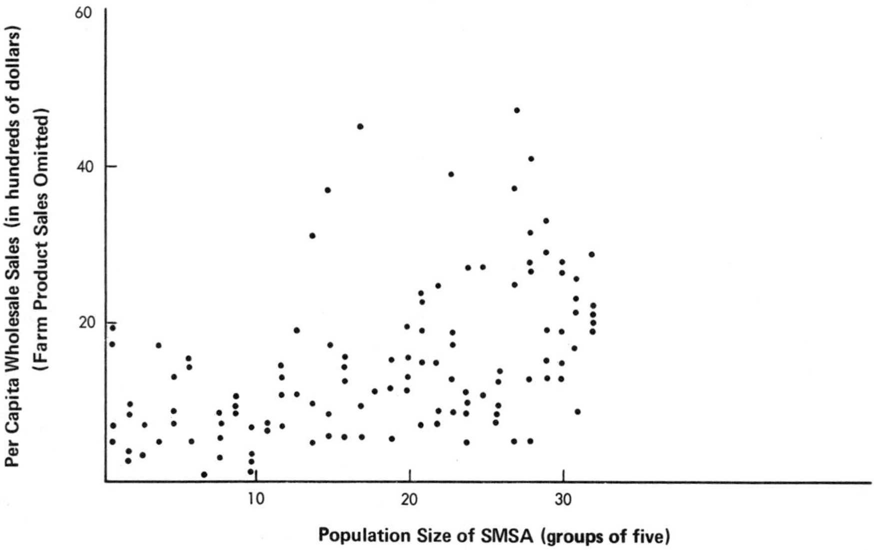

Figure 15

drop in the proportion of the metropolitan population engaged in wholesale trade as we move into successively smaller cities. This assumption grows out of the belief that wholesale trade is directly related to some hierarchical position that the city may hold. When we test the truth, we find that our notion is basically supported but needs some modification.

There is a decrease in relative importance of wholesaling employment in successively smaller cities, moving downward from the largest until the bottom quintile of population is reached, when wholesaling employment tends again to become more important. The explanation is simple and quite logical when we reflect on it.

We have seen that wholesale trade has components of location related to importation (entrepôts) and to arbitration. Additionally it has components related to a process we might term amplification. That process grows out of the critical need for certain types to be near their customers even when those individuals are small in number and generally isolated in location. In these types of wholesaling, area alone is of significance and it will tend to amplify the small demand for goods within a local area until it is economically of sufficient magnitude to bring about the provision of a service in a particular small area.

Perhaps the clearest example is found in automotive parts distribution, an activity in which rapid access of the customer, often a large but undercapitalized group of garage owners, is critical to the wholesaler. This requirement of proximity brings about an amplification of small demand in isolated places to a scale sufficient to support marginal "stockroom" wholesalers widely distributed in space. Pushing the analysis one step farther, we may note that even in fairly densely settled areas these stockroom types of wholesaling exist and, particularly within a complex of cities located close together, the smaller cities will have stockroom wholesaling even though they may have had to yield all other types of trade to the dominating central cities of the urban complex.

Thus, even in metropolitan satellites, such as Brockton or San Jose, there will be amplification, which we may define as a tendency for some wholesaling to reflect conditions of access, to the extent that either in isolated locations or in metropolitan satellites (in some lines) a small demand will be judged as sufficient to encourage the creation of a few service-oriented wholesaling establishments. In complex urban areas this service principle also amplifies weak bases of wholesale trade. The smaller SMSAs in Megalopolis exist in a milieu that calls for rapid and highly elaborated service, so even Stamford or Waterbury will have some stockroom wholesaling.

Per Capita Wholesale Sales. When we relate per capita wholesale sales to the population of respective SMSAs, this tendency for amplification is equally apparent. In Figure 15 (page 122) the scatter is wide, but there is a trend that shows a trough in the fourth quintile from the top and a rise in the lowest quintile, as we would expect under the operation of amplification. This similarity of employment and per capita sales gives us some assurance that amplification does not merely mean the undertaking of submarginal operation in this smallest fifth of the SMSAs; sales grow with the amplification of wholesaling.

Wholesale Establishments in Relation to Population. We may measure the number of wholesale establishments per thousand persons in an SMSA and secure an understanding of structural and areal distinctions. Thus, in Figure 15 the values on the *Y*-axis may be read directly, so that we have a ranking of all cities without regard to the size of the city. When we add in the values of the *X*-axis we split this ranking apart and look to see how city size affects it. Then we can see clearly that *augmentation* is important in explaining the concentration of wholesaling establishments. This notion is founded on the view *that there is a normative value for wholesaling of one to two establishments per thousand population.* If we ask what places have more establishments, the type of cities that stand above the level of two per thousand provides us an answer. They are for the most part those cities that are part of a well-developed nationally based wholesaling structure, or else cities that, for a diversity of reasons, have managed to create a geographical supplement to their normal tributary area. As an example of this geographical augmentation we may look at Portland, Maine. The relationship of Maine to the rest of the United States, and the concentration of transportation to and from the other states, mean that almost all trade for Maine must pass through Portland, allowing that city a degree of domination that exceeds even that of Jacksonville with respect to Florida. Even in Alaska the general use of sea transportation opens more entrepôts than is the case in Maine.

In the group of cities with over two establishments per thousand people, the Missouri River alignment of towns—Kansas City, St. Joseph, Omaha, Sioux City (and by extension) Sioux Falls—all experience such geographical supplementation to a degree. The scale of augmentation enjoyed by the Middle Border towns has been declining rapidly with the growth of production and manufacture in the West, Far West, and Southwest whose products may enter the Great Plains and Mountain areas without passing through the Missouri River alignment. Charlotte, Atlanta, and Dallas represent a variation of this pattern, for they stood at the original railroad entrance points into an extensive monocultural farming area. Trading with the rest of the country was the essence of economic life.

A distinct form of augmentation exists in the cities with specialized trading functions. Boston may suffice as an example. Because of the pioneering of New England in manufacture, wool and leather trading centered in Boston and there were wholesalers, traders in these staples, for whom the United States as a whole was the market. Sioux City, Omaha, St. Joseph, and Kansas City carried on this sort of specialized trading with respect to cattle, while Greensboro–High Point, Fort Smith, and Grand Rapids all garnered much support from furniture wholesaling for the nation as a whole.

In farm and oilfield equipment sales it is common for the manufacturer to control distribution and service through carefully designed franchises. These exclusive distributing contracts facilitate the manufac-

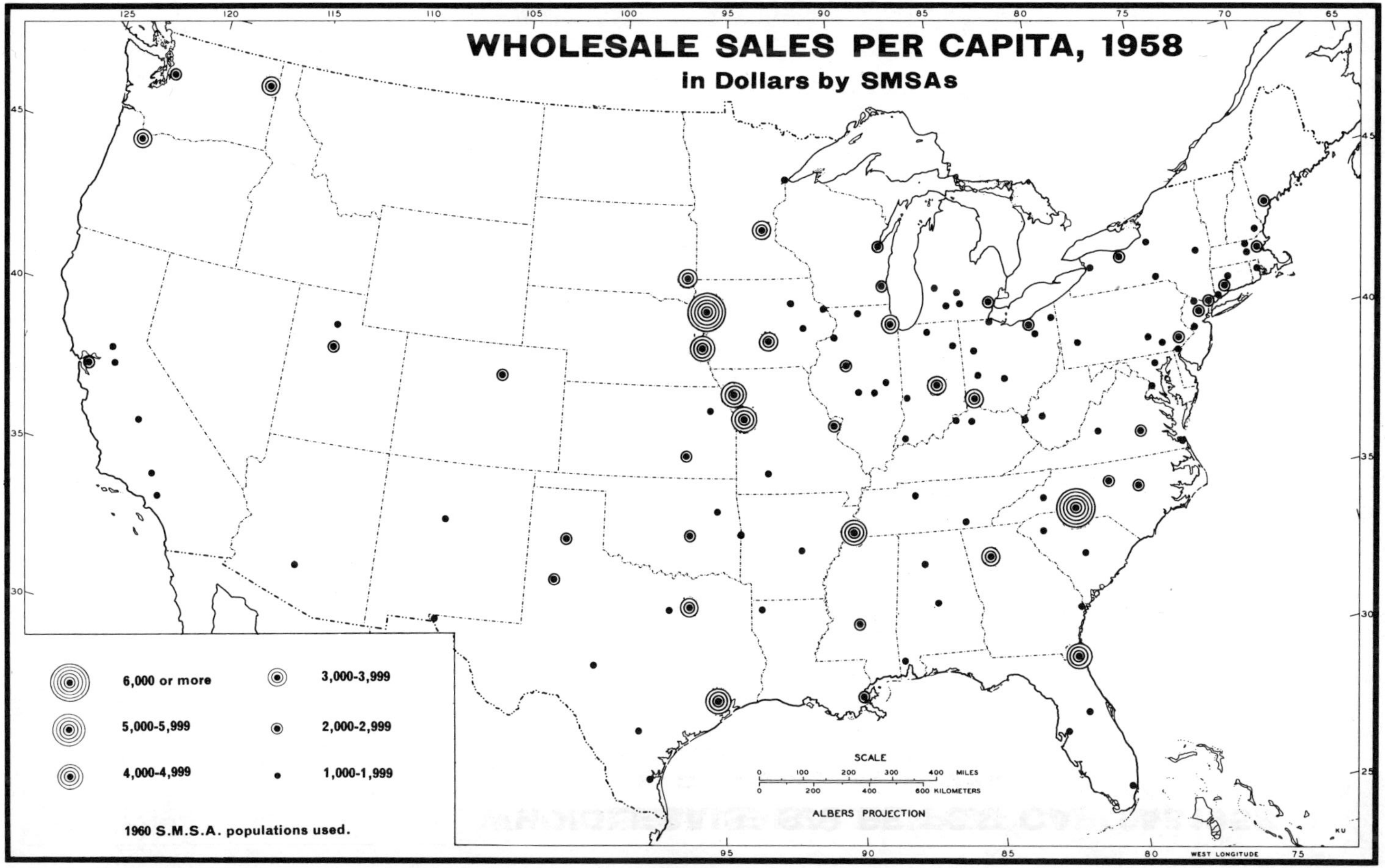
WHOLESALE SALES PER CAPITA, 1958
in Dollars by SMSAs
6,000 or more
5,000-5,999
4,000-4,999
3,000-3,999
2,000-2,999
1,000-1,999
1960 S.M.S.A. populations used.
SCALE
0 100 200 300 400 MILES
0 200 400 600 KILOMETERS
ALBERS PROJECTION
WEST LONGITUDE
KU

Figure 16

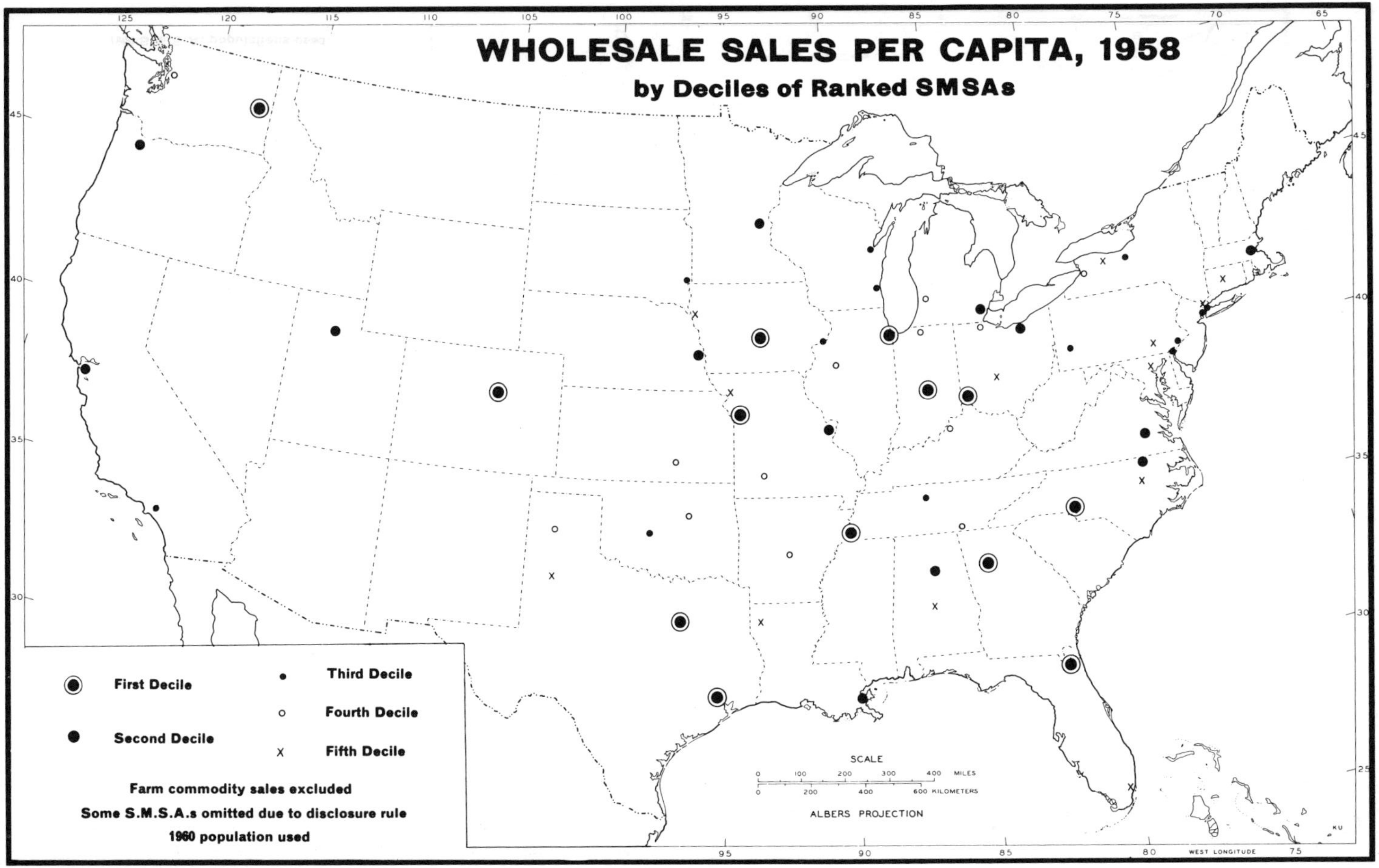
WHOLESALE SALES PER CAPITA, 1958
by Deciles of Ranked SMSAs
First Decile
Second Decile
Third Decile
Fourth Decile
Fifth Decile
Farm commodity sales excluded
Some S.M.S.A.s omitted due to disclosure rule
1960 population used
SCALE
0 100 200 300 400 MILES
0 200 400 600 KILOMETERS
ALBERS PROJECTION
WEST LONGITUDE

Figure 17

turer's control of selling, so the end result is much the same as that sustained when the manufacturer carries on his own wholesaling. Thus, we may lump together franchise selling and controlled sales agencies and branches and note that either type of business organization represents a conscious structuring of the pattern of distribution. The manufacturer, so long as he can manipulate sales in a geographical sense, will attempt to concentrate those sales into as few offices as possible. The simple result is that in areas with franchise selling, the large cities will tend to be overly large, as sellers, and the small cities may live in the shadow of neglect. But, of course, some small cities have their time in the sun as a result of such control. Tulsa and Bartlesville, Okla.; Midland and Odessa, Tex.; and Williston, N. Dak., all show design in their wholesaling structure. Thus, we may allude to this careful planning of the location of sale and its implementation through designated selling territories as the process of *manipulation.*

Conclusions

The location of wholesaling activity in the context of national space poses a problem in synthesis more than analysis. No single conceptual mechanism can account for the diversity of forces at work in shaping that locational pattern for, unlike the situation in retailing, in wholesale trade we deal with repeating and abstracted transactions to the extent that space has a reduced psychological quality and a variable economic impact. No single explanation, such as customer gravitation which provides the cornerstone of central-place theory, can suffice. Wholesaling centers carry out several distinct functions, and wholesale trade can take place several times over before the ultimate retail transaction takes place.

Theories of wholesaling location are synthetic, and we must distinguish the elements in that composition. The historical component can be subsumed in a concern for the continuing process of *internalization,* which causes the trade to be located in ever more direct response to the needs of its customers, rather than its suppliers, as economic development takes places.

Because wholesale trade must, in certain lines, seek to determine national and regional taste, we find the process of *arbitration* bringing certain cities to a prominence that causes buyers to converge upon them. In still different cases, space in its simple form assumes a dominating force in determining wholesaling location. In those instances when proximity becomes important to permit rapid demand satisfaction, local support will rise above marginal values through the process of *amplification,* which takes small units of support and enhances them through increased profit margins arising from the substitution of more critical time utility for greater volume.

The rigidly space determined pattern of wholesaling is modified by the force of *augmentation,* which tends to elevate certain places to an enlarged local wholesaling importance. The common form this takes is the augmentation of a normal trade area due to geometrical or geographical accident. Maine and Florida form augmented market areas for Portland and Jacksonville, respectively, and most of the cities along the Middle Border on the Missouri River experience such enlarged hinterlands in the openness of the Great Plains. Augmentation also comes from exceptional wholesaling such as that associated with a commodity auction market (cattle in Omaha) or local manufacture (Fort Smith or High Point in furniture).

Some of the structure of wholesaling comes from purely external forces and represents the result of the *manipulation* of location by firms with headquarters elsewhere. The result is a pattern of establishments related more to corporate strategy than to regional economic geography. Finally, we may acknowledge that the single most constantly operative force in shaping the pattern of wholesale location is that of *regionalization,* which is always present, even if repressed, in the earlier stages of economic development, and comes into flower after internalization and amplification have had time to accomplish their goals. At that time, but only then, does the pattern of wholesaling begin to show the nesting structure proposed by central-place theory, and it still continues to preserve *and use* relicts of past wholesaling geography that come first as entrepôts, depots, or centers of taste arbitration.

CHAPTER 6 *wholesaling within cities, a brief postcript*

In the format of this discussion of the geography of wholesaling, space forbids a detailed consideration of the location of the trade within cities and metropolitan areas. Yet it seems desirable at least to allude briefly to the bearing that some of the ideas and concepts advanced have on such location problems. Experience has shown that there are two levels of focus in location studies—what we might call the allocation of economic activities in national-or-world-space and the allocation of space within the city to particular economic undertakings. The body of this short book has had to be given to seeking an answer to the location of wholesaling in national-and-world-space. Still, if the thoughts about location put forward have general application, we should be able to make some use of them in explaining where wholesalers will choose to set up shop within the city region.

The most concise application of those notions comes through an aggregative classification of intra-urban wholesale location, which can then be tested against the concepts of location advanced. Such a classification is itself of relatively little intellectual worth, standing neither as an explanation of process nor as the statement of full truth. Taxonomy should exist only as an aid to further understanding and it is that purpose which is here envisaged. A second qualification must be entered as well, that the detailed evidence upon which this taxonomy is based is geographically restricted to the San Francisco-Oakland-San Jose sympolis,[1] though comparison with a number of other urban structures suggests wider support.

[1] In the case of the urban structure that is commonly alluded to as the "Bay Area," that is the San Francisco-Oakland SMSA, and that for San Jose, the use of the morphological concept of the metropolis is inappropriate because it makes the fundamental presumption that there was in the Bay Area a generational spread of the city.

A Wholesaling Taxonomy

THE TRADITIONAL WAREHOUSE. In the strongly historical context of our general discussion it seems appropriate to begin our classification in terms of evolution. In fact, as substantiation for the theoretical assertions put forth, we should be able to explain location of wholesaling within the city in the same historically relative propositions that have been used in the broader consideration. We can start out with the notion that within the city every settlement has a *point of inception,* where the town is attached to a larger structural system. Taking that point of attachment as given, the initial wholesaling district will depend, in specific location, *upon the particular nature of the original transportation medium.* In a railroad town the warehouses will wall in the freight yards, in a port the wharfhead warehouse will dominate, and in a town growing as a distribution center in the era of motor trucking, the city outskirts adjacent to the main long-distance artery will become the fledgling wholesale district. The process in each case is the same, even if the actual location differs and the actual shape of the buildings varies because of contrasts in the technology typical of the time when it was built. In the port, which is likely to be the oldest within our context of these districts, the emphasis is probably on vertical movement simply because man learned to use the pulley block-and-tackle earlier than he learned other means of goods-handling. The vertical handling was well developed on the sailing ship and it came quickly to characterize the movement of stores in contemporary quayside buildings.

The railroad was obviously more subject to ramification than was the quay; thus railroad wholesaling could be more dispersed. Furthermore, it was more horizontally separable, to the extent that the boxcars could be shifted from place to place along the margin of the building and could be successively spotted at a number of unloading points. Yet within the building, vertical technology predominated. The morphological outcome was greater scale for warehouses, with an increased emphasis on peripheral access; however, technology remained important, as shown by the continuing construction of multistory warehouses.

When motor trucking was introduced, the vehicle itself came to be used to a considerable degree, even within the building, where it could be employed to move smaller loads from place to place simply because of its modular size (perhaps ten tons rather than the thirty to forty tons for a boxcar). Soon after the development of trucks came the forklift-pallet technology for the handling of goods. The combination of the truck and forklift-pallet so completely overpowered the vertical handling of materials that the single-story warehouse became the norm after

---

As I have attempted to assert elsewhere [*Geography and Urban Evolution in the San Francisco Bay Area,* 1964 (Berkeley: Institute of Governmental Studies, University of California)], some cities grow with all parts as siblings, and in such a situation it seems more accurate to emphasize the concerting of growth, as the term sympolis does, rather than the outward spread from an older core, as the term metropolis does.

about 1940, and its location was largely controlled by highway access, even though a number of paleo-trucking warehouses were built on spur-tracks, many of which stand unused today. Their builders failed fully to understand the real nature of the system they had adopted, and lacked the courage to break with the past.

Thus, in every major city there was a "traditional wholesaling district" whose morphology depended on the time of inception and the conditions then controlling, but whose site would be adjacent to the core of the city and within the central business district "frame" that Horwood and Boyce defined. In truth the wholesaling district became one of the major physical determinants of the shape of the central business district and tended ultimately to inherit the original site of the retail district as that unit grew, abandoning its outdated original accommodation.[2]

At this stage we may anticipate our story a bit when we point out that the traditional wholesaling district grew up and was sited mainly in response to the *conditions of access from the sources of supply.* Thus, the district shared the rôle played by both the points of initiation (of trade) within the mother country and the early entrepôts in the new lands settled under the mercantile model. We should fully understand that the conditions of access *within* the city were not the main determinant of location; rather the wholesaler set up shop at a *specific point* in an areally extensive distribution system which happened to call for a precise location within the city largely in terms of the contemporary transportation technology. This was particularly true because the wholesaling establishment was more likely to be an articulation along the course of a complex linkage (between producer and ultimate consumer) than the end itself. Thus, its site was chosen within a field of possible location narrowly defined by current technology. That area of choice could be greatly enlarged when there were improvements in transport technology.

DISTRICTS OF CUSTOMER ACCESS

The Produce District. At the core of the city a second general class of wholesaling establishments was well-developed in the past, though much less so today. These were the *establishments whose location was predicated upon customer access.* Perhaps the clearest example is found in the traditional *produce district* wherein often-numerous small producers, or their agents, met consistently-numerous small consumers drawn from a city's restaurants, hotels, and vegetable dealers. In such a context we should not be surprised that produce districts were normally at the very edge of the shopping district, often in areas susceptible for inclusion in an expanding office quarter. The result has been ever increasingly difficult access for both producers and consumers, who are no longer so likely to reside mainly in the heart of the city. These im-

[2] For a more full discussion of the process of take-over see R. E. Murphy and J. E. Vance, Jr., "Delimiting the CBD," *Economic Geography,* 30 (July 1954).

pediments to access have combined with the obvious attractions of the area as "development land" to lead to the casting of successfully covetous glances, on produce districts, by the practitioners of urban renewal. Thus, these market quarters are fast disappearing from our larger cities.

Product Comparison Districts. Equally influenced in the beginning by customer conflux were the premises of a number of wholesalers for whom comparison of goods became important. Such *product-comparison districts* were particularly characteristic of the clothing and furniture trades. Garment districts were not merely the clustering of manufactures to secure agglomeration economies, they also became a way of selling. For that reason they were often set up adjacent to the part of the central business district where out-of-town buyers could come together in hotels. Similar locations were chosen for furniture marts but they, of obvious need, had to stand as selling agencies and did not house factories as well. Consequently furniture has tended to have its wholesaling concentrated in a single or a few adjacent buildings (as in Chicago, San Francisco, Grand Rapids, and High Point) where externally located manufactures bring together their goods for the convenience of customers who come periodically to the mart city to select goods by observation of the full-range of choice.

Will-Call Delivery District. A third form of wholesaling related to customer conflux shows up in the central location of certain establishments to which small retailers operating within the metropolis come in order to secure immediate delivery of items that have timeliness. The clearest example is perhaps found in the phonograph record business wherein violently mercurial "taste" governs the salability of an item. No retailer wishes to buy too heavily for fear that the wind will veer, leaving him with a counter covered with last week's formula for success. In such a situation the ability to go to "the city" to secure a new supply of the latest hit is important. There is no need for will-call delivery wholesalers to be near the hotels, or other facilities geared to the wants of out-of-town buyers, but there is the need to be handy to the internal metropolitan highway system. Similar location conditions affect certain kinds of appliance, plumbing fixture, auto parts, and drug wholesaling wherein the wholesale merchant undertakes to provide large-scale stockroom wholesaling for a numerous group of often undercapitalized retailers who operate in small premises spread about the metropolis.

Manufacturing Stocks District. Still another type of wholesaling in which customer access weighs heavily is the selling of paper for the printing trades. Because this is a relatively standardized supply item, for an highly diversified manufacturing process, which through its tie to commercial and office activities locates near the city core, we tend to find paper wholesalers near the city core and usually near the printing quarter. Because paper wholesaling is subject to far more economies of scale than is printing, the paper dealers are often very large when their customers may be small.

One of the less common but still more striking manufacturer's stock wholesaling districts is associated with the garment trade. Because the garment trades are characterized by many small producers, a considerable scarcity of working capital, violent shifts in fashion with consequent changes in the needs for "findings" and other raw materials of production, and a great time pressure on production, it is common to find an intimate symbiosis between garment makers and the wholesalers supplying their raw materials. The midtown district in New York and the south Main Street area in Los Angeles are clear examples of adjacent wholesalers and manufacturers. Locationally the manufacturer, who is his own wholesaler, chooses a site near the periodic pied-à-terre of his customers for his plant and his own wholesalers of findings must take up residence there.

The Office Wholesaling District. There is not sufficient space here to deal with all the types of wholesaling strongly influenced by the conflux of customers. But it is critical to emphasize that today great importance is attached to the *distinction between customer access and supply-of-goods access.* If supply-of-goods dominates the location decision, then technological changes will have greatest affect on the location decision, and we may find during the course of the city's history several rather different sites that have been used by a single wholesaling firm, with one toward the periphery standing as the present home. In contrast, when customer access dominates, locational shift will be less common, for the simple reason that customers and their transport have changed less in the course of urban history than has the handling of goods. And given the annular growth typical of most metropolises, the core area is likely to remain the point of maximum collective access for customers whose own business location is as widespread as the urban population. For those buyers coming from beyond the edge of the metropolis a financial-office district location for the wholesaler of the title to goods is even more a point of minimum modal access.

In this context the persistence within the office district of various types of wholesalers dealing in the *title* to goods rather than their physical handling is intelligible. For these agents, brokers, and the like, the location problem becomes one of reaching men, and often agreement as to the market price. In that rôle the medieval face-to-face dealings of the office district are at least helpful and often essential.

The Decline of Customer Conflux and the Increase of Transport Costs

Great shifts are taking place in the location of wholesaling establishments. In simple truth, physical access to the customer has declined in importance due (*1*) to an increasing standardization of goods, (*2*) to a greater ease in dealing with customers at a distance through electronic communication, (*3*) to the ability to set prices in isolation rather than surrounded by the goods involved, and (*4*) to the startlingly increasing bite of transportation in the establishment of the final price for goods. Taking the last of these matters as the only one needing amplification, it can be shown that today the cost of transport for goods forms such a major component of the final cost that the chance to lower the unit F.O.B. cost of the good through core-area selling may be fully countervailed by savings in transport charges to be gained from distribution from peripheral warehouse location. The problem is clearly a very complex one and not amenable to analysis in this restricted appraisal. However, it should not be forgotten that in the present state of economic geography the transport component tends to gain weight at the expense of scale and agglomeration economies and "centrality."

The peripheral shift of wholesaling stems from an increasing weighting of transport and the technology of goods-handling in the establishment of wholesale prices. The best example is found in the handling of food, the most bulky item in wide use and the one wherein time-values make transport of critical importance. Although it may be cheaper to ship carloads or truckloads of lettuce or meat, as it is of canned goods, the concern for time of shipment forfends such scale economies. When transport costs cannot be minimized through demand accumulation and infrequent shipping, they must be minimized through the selection of a location wherein relatively frequent transport movements can be carried on at the least cost. In its essential form this calls for a reduction of terminal costs, through modern goods-handling processes, but much more it means reducing the line costs of actual trucking. Without entering into that proposition in detail, it seems valid to state summarily that peripheral freeway-oriented sites afford such line cost minimization to the highest degree because each hour of the driver's time accomplishes the most line transport. In contrast, for those operations where the wholesaler does not have to take account of line costs—that is, basically, where the customer does the line transportation—the interest will center on trying to secure the most widespread reduction in line costs for the *individual* customer, so as to create the greatest impression of cheapness for the greatest number of customers. In such a situation the minimum modal distance tends to rule in the location decision and often causes operations in those lines of trade with large intrametropolitan groups of customers to remain in an old core site.

OUTLYING WHOLESALING. The outlying wholesaling districts tend to be used most by wholesalers who provide the transport and final delivery to the retailer, as in petroleum, food, automobiles, and trucks, and other widely consumed and highly describable goods. Several rather exceptional lines appear in outlying areas. There is what we might term mail-order wholesaling, for which the use of postal distribution is envisaged. Perhaps the best example is the general line of presciption pharmaceuticals. Given the large number of small pharmacies and the wide spread of physicians, mail shipment of pharmaceuticals is common. With the zone rates on parcel post, or even the nonzonal quality of first-class mail, a suburban post office may afford the pharmaceutical wholesaler an easier and more pleasant location in which to work than one adjacent to the overloaded central city post-office.

The final outlying wholesaling type we should consider is the small-scale stockroom operation. Already we have noted the tendency, even within the city core, for certain wholesale establishments to cater to a metropolitan-wide market calling for the holding of stocks that aid small-capital retailers. Still more directly related to this stock-holding need is the existence (in most outlying subdivisions of the metropolis) of wholesalers of parts and small or very standardized items to supply various service trades. The lumberyard, serving mainly the individual carpenter-builder, is the obvious example but we might also mention auto-parts dealers, plumbing fixture dealers, and electrical supply houses. In each case the workman's time is more valuable in pursuit of his trade than in pursuit of his raw materials so the wholesaler must locate so as to minimize the workman's journey, not to create some minimum modal travel pattern such as that found in stockroom operations at the city-core. The retailer uses his time in selling, and thus he can justify the trip to the edge of the central business district, but the skilled workman uses his time in making things, and thus he can justify only short trips to the outlying stockroom wholesaler.

The Link Between Urban and National Wholesaling Location

Throughout this discussion we have come back again and again to the notion that an over-all structure *specific to wholesaling* shapes the pattern of trade and the location of its establishments. The mercantile model, discussed in greater detail in the chapter which follows, sets out to explain, on the basis of the historical facts presented so far, the physical expression of that economic design of trade, which is mercantilism and its conceptual heirs. It should be patent by this point that the specific structure of wholesaling, as the specific structures of other functional components of urban activity, evolves over time. Only by predicating our analysis on a dynamic view of location can we deal with actual observations rather than mental abstractions. Within the city there is a richness of observation which particularly calls for a dynamic analysis and our broad nationally-based model of wholesaling location can answer that call.

The evolution in urban wholesale location conforms to the broad proposals drawn from national patterns. Within the city, as within the national space, trade began at a point of initiation, be it a coastal entrepôt for all of North America or the warehouse next to the railroad depot in Omaha or Winnipeg. From such a common hearth the structure of trade *specific to the geographical entity* evolved, and we cannot understand its mature form without understanding its infant form. As was true in the national pattern, within the city certain of those establishments which grew up at the hearth never left its warmth. The option that some types of trade had to stay in the coastal entrepôt or near the point of initiation in the town came from their opportunity to use fixed trading ties (agency of trade) to expand the outer edge of the trading territory without moving its center. So long as the costs of delivery and transport could be kept down, this maintenance of core location offered enhanced competition, through the opportunity to expand service in an enlarging market. But in time the friction of space came to weigh more heavily as was signalled by an increasing contrast between the efficiency of transport carried on in the extrametropolitan region and the worsening situation within the metropolis itself. At the long-distance linkages grew, the advantage of core location disappeared, leading ultimately to a peripheral location decision. Such a shift *within the city* is a direct response to the external orientation of wholesale trade, and in keeping with the national experience in wholesaling.

The application of the notions of arbiter-markets and price-setting cities is as clear internal to the city as in the nation as a whole. The fundamental control has been that of access for persons, and there can be little question that city centers have that quality to the highest degree. In fact, the largest cities in the past have tended to have the attribute to the largest degree. But taking a specific case, the awareness of style grows out of knowledge of the products in the market and the customers' response to those items. New York has set taste throughout most of the period of trading in women's ready-made clothing. Now, however, access to New York is easy only in the abstract—airline schedules head toward New York with greater frequency than anywhere else. Actually landing, reaching the center of Manhattan, and trailing among the many showrooms today could produce a notion of efficiency only in the mind of the most perverse. Thus, there is considerable significance to be attached to the fact that starting in 1964 an outlying Apparel Mart built in Dallas began seriously to compete with midtown Manhattan in the arbitration of taste. The president of one of the largest women's fashion-clothing manufacturers, operating in Manhattan, said of his Dallas mart, "Here's where we get our trends for the season." As the *New York Times* noted "The trends are charted by the 10,000 store buyers who drive from neighboring states or fly from as far as Canada to get what amounts to an instant view of fashion developments in everything from shoes to evening dresses. Smaller stores are no longer dependent on the visits of individual salesmen to their particular towns to know what is going on, and larger ones do not have to get to both the East and West Coasts." [3] Getting to Dallas from much of North America is quicker than getting to New York and, once there, buyers need only visit a single modern building rather than the diffuse chaos of midtown Manhattan.

Within the city this ability to bring together those whose direct responsibility rests in the creation of a "market" maintains furniture and fashion quarters, commodity markets, and the purlieus of agents and brokers in general. Because commodity trading demands a broad range of financial services, it tends to be carried on next to the financial district, as in the Board of Trade Building in Chicago or in the Winnipeg Grain Exchange; and it can only be carried on in a very few cities, where one

[3] Bernadine Morris, "Texas Stakes Claim to Fashion Honors," *New York Times*, 29 October 1969. © 1969 by The New York Times Company. Reprinted by permission.

finds the groupings of men necessary to set prices. In this situation we find that evolution has its limits; Winnipeg might replace Chicago, or Dallas, Manhattan, but the physical structure within the city is likely to remain unchanged. Even the outward movement of *Les Halles* to a new produce district on the edge of Paris has not changed the form of that market, which remains in broad morphology very much as it was when rebuilt in 1854, which in turn was mainly an enlargement of a market started in 1183.

This brief appraisal of the intrametropolitan location of wholesaling establishments has suggested that analysis must be undertaken within a framework which is far from parochial, far from restricted to the closed system of central-place theory. Both in the world at large and in the urban microcosm an outward-looking and evolutionary model of economic geographic location is called for, and as our final undertaking the last chapter will be given over to its proposal.

CHAPTER 7 *wholesaling: the geography of enterprise*

In contemplating patterns of trade and the geographical system of mercantile activity, one must ultimately search for an understanding of a general design. Understanding that design requires analysis into components, but the utility of the design rests on the integration of those components into a functional model. Geography as a field of study has only in the present generation perceived the need for the development of such models. Our intellectual ancestors thought themselves explorers of an unknown world, whose product was factual knowledge and for whom success meant collecting even greater knowledge. No longer does the unknown have much importance: the unexplored thickets hide little but trivial truth. The contempory challenge is the *true comprehension* of the facts already available. In this concluding chapter I wish to reflect upon the "facts" of wholesaling in a search for the functional model of activity that they suggest.

The Geography of Enterprise and Historical Economic Geography

Throughout this appraisal of wholesaling the strongest thread in the fabric has been that of enterprise, originally individual and ultimately corporate. No student of location should for a moment doubt the role that the entrepreneur plays in the creation of systems of activity that ultimately become the grist for geographical analysis. Yet, only recently has much attention been turned toward distinguishing the entrepreneurial contribution, and we must seek to understand this inattention over such a long period.

NATURAL ORDER. When geographers began to recognize the rapidly diminishing return to be secured from the worn-out ground of classical field work, the search for mental fertilizer produced Darwinian notions of *natural order*. Simply stated, these ideas held that there is a coherent and repeating design in the nature of geographical occurrence. If we discern any part of that design, we may sketch in the unobserved details in much the way museologists fill in the missing shards in an Etruscan vase. The operational utility of a natural-order approach, such as that implied in central-place theory, is obvious. At the same time, the mechanistic quality of that approach often relegates both space and the acts of men to the status of constants variable only through aggregation. In such a context the entrepreneur is assigned no role, and human behavior becomes a delinquent to be frowned on as antipathetic to law and order.

EXPERIENCE. To introduce the merchant and give him a purpose we must turn away from any predestined approach, seeking instead the exegetic account of historical events. Through such a shift, different outcomes from similar beginnings become possible, and in place of mechanistic "order" with fixed responses we encounter experience with its normal parameters. This range of experience is one of the fundamental virtues of an historical system in that such a range envisages an over-all scalar limit to activity, but a large practical differential in the assignment of roles to various points and persons within the system. The notion is that of historical and cultural relativity wherein man plays a part in accepting or rejecting the natural forces. In this way there is no total break between the natural-order school of thought and that of experience, but it would be foolish to hold that the two are free of antithetical qualities.

Within the natural-order approach all is revealed or assumed at the beginning, and deduction is the main intellectual process. The experiential approach argues instead that acts have an historical context which must be perceived and analyzed if one is to understand the meaning of present events. Thus, induction is the process most fundamental to the historical approach. The specific application here appears as historical economic geography wherein the rational explanation of a system of activity requires concern not merely with present distributions but also with past patterns whose structure gave a form to the system that has, even in being modified, persisted to the present.

The importance of this distinction between natural order and experience, as the explanation of geographical structure, becomes manifest when we return to the part the merchant played in the shaping of urban structure. The very notion of the entrepreneur carries with it assumptions of risk and uncertainty on his part. Such gambles stem from the exigencies of locational decisions that he must make to gain profit and success, in sharp contrast to the mechanistic regularity of location and scale implied under central-place theory. Furthermore, the notion of personal enterprise argues that individuals may shape different *organizations* of space within an economic frame and thereby focus trade and its profit on several quite genuinely alternative places. Thus, *man,* and the history of his actions, becomes an agent as well as a victim of change in a spatial order.

It would be incorrect to hold that no regularity can be anticipated in the experiential approach to geographical analysis. Human behavior is far from being unstructured and quixotic. It is subject to shared responses to common situations. At the same time, individual solutions to situations can be expected, and the study of the experience of man in shaping resolutions for problems has two logical concerns: that with the *commonly shared responses* and that with the *exceptional acts*. In the latter class fall the instances of striking entrepreneurial action wherein an individual or a firm seeks, by departure from regularity, to shape an organization of business that will gather greater individual, as opposed to generic, profits. Such is the rationale of capitalism, and it is indeed striking that the standard theory of spatial organization that we as geographers apply is so very unresponsive to that objective. Central-place theory is essentially bureaucratic in conception and unfriendly to innovation. Its very units of measurement are those of a rigid Germanic political order, and its warmest acceptance is to be found among the architects of planned or command economies. This may be the rationale of a statist future, but it is a poor exegesis on a mercantile and capitalist past. As geographers, we regard our basic concern to be the discovery and understanding of how we reached the present more than to predict the events to come. If we continue in that role, it seems both logical and intelligent to try to find a better analytical device than that afforded by central-place theory.

By looking at *both* the regularities of response to past situations and the departure therefrom, we may hope to gain a better understanding. As we have seen, the structure of wholesaling offers us both the reason and the raw material from which to shape a new form of analysis, and as it will subsequently appear, a new model of settlement structure.

The Historical Basis of the Central-place Theory of Settlement

A critical appraisal of central-place theory suggests several fundamental characteristics of the system:

1. Its existence is an outgrowth of an areally based support within which relatively consistent spatial economies must be assumed.
2. "Retail gravitation" is the main force shaping settlement patterns.
3. This gravitation is shaped by customer convenience, with distance built into the system in terms of a time-and-cost factor that determines decisions customers make about the "central place" they will visit.
4. Expansion in the system comes from growth in consumption, rather than production. Thus, growth is dispersed and must be focused through retail gravitation on a particular place to bring about change.
5. In central-place theory there is the basic assumption that the customer

goes to the central place and, for that reason, access to goods and services can be seen in terms of the willingness of persons to move over a certain distance. Willingness is measured by distance not time.

6. In this context settlement begins with customer conflux, and there is the assumption that places of greater relative importance require a greater amount of conflux, creating a hierarchy of places in terms of the distance over which they can draw customers.
7. Because higher relative status among central places is predicated on more units of consumption (therefore of greater centrality of location within a prospective tributary area), we expect that the larger market areas must result from hierarchical aggregation of many basic low-order trading territories and must show even and geometrically centered spacing of trade centers of a particular rank order.
8. In this context a hierarchial foundation must be present to account for the existence of higher-order central places. In turn, those places will be widely and evenly spaced in geographical occurrence.
9. This system is evolutionary from the smallest to the largest components and requires the growth of population and personal income to bring about change.
10. That change can only be expressed initially at the bottom of the hierarchy and from internalized characteristics. Thus, central-place theory is highly endogenic.
11. The expression of change is mechanistic rather than variable, and it does not comprehend differences in human behavior or social objectives, either existing or historical.
12. Thus, the role of the entrepreneur is missing from central-place theory.
13. Furthermore, the central-place structure is predicated upon a geographically closed system with the constraints it introduces. Notable among these is the inhibition of innovation in goods and demands.
14. Yet History tells us that as early as the Middle Ages experience did not so constrain trade, even within the retailing structure. Innovation in selling was often introduced *exogenically*.[1]
15. Given the constraint of the closed geographical system, it is difficult to explain why local customers would seek to secure goods of which they were unaware in response to cravings they would be unlikely to make conscious.
16. Once we admit of exogenic stimulation of *retail* trade, the closed system no longer holds, the role of the entrepreneur is introduced, the ability to forge long-distance trading ties maintained through the work of agents is foreseen, and the basic and *necessary* constraining assumptions of central-place theory are relaxed.
17. Only if we may parochialize trade through the operation of restraints,

[1] Fritz Rörig in his study of *The Medieval Town* (in translation, Berkeley: University of California Press, 1967) shows clearly that when printing was invented it was essential for "such a large-scale business as that of Antonius Koberger the publisher . . . [to have] agents in Paris and Lyons, who at the same time systematically sent out men who had been specially schooled in the job of retailing the published works" (p. 147). In this they merely perpetuated the facilitation of *retailing* in foreign and innovative goods represented by "the 'cutting rights' of the valuable Flemish materials held by the leading men of the wholesale and foreign trade in the German towns which lasted until a circle of local retailers having a specialised knowledge, and devoting itself to the retailing of Flemish cloths, was formed, which allowed the wholesale traders progressively to renounce their 'cutting rights'" (pp. 146–47). [emphasis supplied]

such as those of feudalism or present-day "command economies," may we counteract the natural tendency to universalize trade.

18. It is not illogical, or against actual experience in some places, to argue for a central-place model. It is, however, both unjustified and unsupported to argue for the model as a timeless universal rather than an historically relevant *special case*.
19. If central-place theory explains only a special case, it is obvious that there are additional explanations of trade and town structure, and we should seek to discover them.
20. Thus it has been essential in this study to examine constantly and critically the propositions of central-place theory and to note their limitations. The purpose is not to convert believers so much as to broaden their outlook: a reformation rather than a conversion is our objective.

With these characteristics in mind we may distinguish several lines of weakness (as an explanation for the geographical pattern of wholesale trade) that appear in central-place theory. Those lines of weakness may be seen in areas other than wholesaling, but space here rules out a full critique of Christaller's theory.

Proceeding through the central-place model we follow a firmly normative quarry, that of small economic geographical units assembled through rigid aggregation into a structure of mechanical consistency with its fellows. Man, and his experience, is rendered so constant a force as to be disregarded, and enterprise and history appear to be of no importance. Similar regimentation of geography is undertaken through an intellectual flattening of terrain and painting over of physical contrasts. In this embalmed world, order exists at the expense of vitality: man must conform or be emasculated into a constant.

These arguments are not merely didactic, in light of the fact that much of the world does not appear to be organized along the lines discerned by Christaller in Bavaria in the early 1930's. It has been difficult enough to find his pattern in the realms of retailing and service industries outside south Germany, but even more, as we have seen here, the central-place model is not an explanation of wholesaling activity. Before seeking to formulate a more encompassing and responsive model, we might well attempt to account for the weaknesses of the central-place design.

Careful analysis of the central-place model leads directly to the conclusion that the brick from which that structure is built is the local trading area with its lowest-order central place (Christaller's "*H*" or more properly "*M*" places). These lowest-order central places (*M*) have a radius of trade of nearly 2.5 miles (an area of 16.7 square miles) in south Germany, a scale similar to that which delimited the smaller feudal holdings of the Middle Ages. These manors or parishes equate to the *M*-place regions and can be assembled into baronies and ever more elevated fiefdoms that correspond in part to the designated hierarchy of German feudalism that was the *Heerschildordnung*. Feudal society and its economic life worked within these cells, and it should not surprise us that the cast of that honeycomb can still be seen in the settlement pattern and conflux of customers in south Germany and other areas developed under the benefice of the feudal order. That we should think of this as the basis for a universal order is, however, astonishing.

If we accept the Marxian view of historical progression even in part, we must grant that feudalism was a distinct economic system that came before capitalism. The onset of mercantilist capitalism brought with it the acceptance of individual enterprise and the abandonment of the ordered society within its honeycomb, which had been the economic model for feudal times. The entrepreneur looked upon space as an ally, not as an enemy; if he could dominate more space, his endeavor would be more adequately rewarded. The traditional and normally immutable order of the Christaller (feudal) conception of spatial organization would have little appeal for the merchant who sought the practice of trade as a way to wealth. In seeking to move away from the medieval practices he found it necessary to turn outward from Europe and to devise a new economic system which Adam Smith quite rightly saw as the supplanting of the much older system of "agriculturalism." This new "mercantilism" became the catechism of wholesale trade and it furnishes us with the force which worked out a settlement model that shaped the new and nonfeudal lands.

Central-place Theory: The Classic "Special Case"

Christaller accepts the fact that his analysis is simplistic, admitting that "we have disregarded, above all, the entire complex of the determination of urban development through industrialization. . . . We have been satisfied with developing the primary net of central places, on which the phenomena of industrialization may be superimposed." [2] To the extent that industrialization is such a superimposition, his method has validity, but in those instances where the industrial revolution shaped new cities and distinctive urban forms, central-place theory is indefensible in terms other than those of a special case.[3]

Mercantilist Capitalism Shares the Atlantic World

The present study of wholesaling, through its concern for the location of large-scale domestic and foreign trade, carried on with the use

[2] *The Central Places of South Germany,* p. 198. Yet as Rörig has shown in history the artisan often sought a quite different and much larger market than that effectively dominated by any central place. "If the shopkeeper was the sworn enemy of free trade for foreign merchants in his own town, and would have liked to see their wholesale selling limited, things were different for the [manufacturing] workers. . . . They needed external trade ties if their livelihood were to grow." Rörig, *op. cit.,* p. 147.

[3] Elsewhere I have shown that industrialization brought sharp endogenous changes in urban morphology, but here it should be reiterated that that internal transformation was in response to the dynamics of housing generalization and segmental labor employment rather than any measure of retail-service gravitation. See "Housing the Worker," *Economic Geography,* 42 (1966): 294–325 and 43 (1967): 95–127. And the scale of manufacturing usually came from external ties which might contrast strongly with the magnitude of trade within the closed central-place system.

of mercantile *agency*, has brought out a further restriction of the application of Christaller's theory. Although in south Germany it may be tentatively possible to hold that central-place dynamics sketch the original settlement pattern (on which later distant trading was superimposed), such an assertion in the Atlantic world, which employed mercantilist capitalism as an economic policy and system, starting in the sixteenth century, is unreal and untenable. Bristol, St. Malo, La Rochelle, Lübeck, and Cadiz, to name only the clearer controverting cases in Europe, and most of the North American entrepôts, cannot be accounted for by Christaller's feudal-economic method. In simple fact, that failure comes not from his use of "economic" argumentation, as he believes, but rather from the narrow focus of his reasoning on a highly restricted and historically relative economic system. Christaller argues that "Depending upon how narrowly one construes the term *economic system*, one obtains theories of greater or lesser individuality. Perhaps, if one interprets economic system to mean 'western,' or more particularly, 'western-capitalistic,' one excludes the economic forms of the feudal Middle Ages." [4] But that exclusion is precisely what he fails to accomplish, because his closed trading system could only be found in a feudal rather than a mercantile-economic form. In fact, I wish here to argue that Christaller's model is so intimately tied up with just those "economic forms of the feudal Middle Ages" that it is unwise to employ his analysis beyond those areas in which feudalism was the economic organization under which major components of the settlement pattern were created. In such usage central place is, in truth, "special theory."

To understand the special theory quality of his analysis it is necessary to look at the contrast between the settlement pattern growing out of a system of agriculturalism and that shaped by mercantilism. To that end we must begin by rejecting the ambiguities of Christaller's study. In his work he notes that "such a division into general theory on the one hand, and special capitalistic theory on the other, has not been made; such a division would be impracticable within the framework of this work. [Furthermore,] Here, the theory should not be systematic, but rather preparatory; it is supposed to serve a particular aim, *to determine geographic reality at present*, that is, to explain the number, sizes, and distribution of central places in southern Germany." [5]

Examining this objective in a critical fashion, it is necessary to point out that only the qualification that *Central Places in Southern Germany* seeks "a particular aim" rather than a general one saves the proposal from total inconsistency. It can be shown (1) that in any settlement situation the massive inheritance of the furniture of the past, along with the fact that a city is man's largest and most durable artifact, makes a topical restriction of attention to the present impossible, except within the abstraction of a natural-order mechanism, which Christaller eschews through his call "to determine geographical reality at present," and (2) that south Germany is such a classical area of the *Heerschildordnung*

[4] Christaller, *op. cit.*, p. 133.
[5] Ibid., pp. 133–34. [emphasis supplied]

that, in any study with such a fundament, it is self-delusion to argue that "one excludes the economic forms of the feudal Middle Ages."

AGRICULTURAL TOWNS. In addition, within the type location for central-place studies, in south Germany, the agricultural "town" is common. Places like Dinkelsbühl and Nördlingen, though part of the classic Christaller pattern, grew up basically as "nucleations" for strength, walled for safety, and dispersed so the resident agricultural workers might live near the fields they tilled. In the central-place analytical model they appear as points of *central* rather than *dispersed* activities (as shown on Christaller's maps of central places and their *M*-regions in south Germany),[6] but historically and physically they were clusters of farmsteads, that is, of dispersed activity. Only derivatively have these farmstead clusters become central places, in a manner true of any sizable grouping of people for whom the local provision of goods and services is both efficient and logical. But the assertion that Dinkelsbühl is there because of a large pattern of retail gravitation is vigorously contested by the simple morphology of the place. Quite clearly there can be no doubt as to the wisdom of Christaller's dictum that "we cannot disregard the historical method, either for special investigations of monographical character, or for general investigations in which the historical method only enables us to understand the ideas of an economy and the concrete economic system born out of it, as a rule, at the time of the creation of the towns."[7]

Consideration of the dynamics of central place, as cited above, strongly supports the notion that this is "special theory" derived from a specific economic system within which two pillars are visible: the parochial economic base of agriculture and the hierarchy-creating political-military base of feudalism (*Heerschildordnung*).

If we characterize the empirical basis for central place as feudal-agricultural, it is ironic indeed that Christaller thought the system free of such a bias. In discussing the contrast between special and general economic theory, in a division wherein he viewed central-place theory as special theory, he notes that "The rules and laws which are formulated in the general theory must, therefore, be valid at all times and in all places regardless of the concrete time and local distinctions of economic systems. . . . The special theory is different; it develops the rules and laws which are valid only for a specific economic system, determined with respect to time and culture. Depending upon how narrowly one construes the term *economic system* one obtains theories of greater or lesser individuality."[8] Instead central-place theory "should not be systematic, but rather preparatory; it is supposed to serve a particular aim, to determine geographic reality at present."[9] Yet how can we pretend that reality has only the present to contend with or that a settlement pattern clearly growing out of feudal agriculturalism can be assumed to

[6] Ibid., pp. 224–25 (Map 4).
[7] Ibid., p. 200.
[8] Ibid., p. 133.
[9] Ibid., pp. 133–34.

be the product of the "geographic reality of the present"? Is it either rational or sufficient to discuss "The *P*-System of Augsburg" [10] as if the decisions of the German Woolworth company were more important than those made by Augsburg's famous Jakob Fugger? There was no doubt in Fugger's own mind that his reach was far greater in extent than merely a commercial appeal to the local Bavarian peasantry, a fact he conveyed in a letter to Charles V in 1523: "It is an established fact that Your Majesty would not have attained the Roman Crown without my help." [11] The commerce of Augsburg already extended to Venice in the fifteenth century and, with the opening of the New World by the Spanish Hapsburgs in the sixteenth century, that commerce reached central and South America. On such a basis, rather than the trading of turnips, was Augsburg to become one of the world's largest cities in 1600 and the Fuggers perhaps Europe's richest men. We should not be misled by its inland location to forget that it was the Welsers from Augsburg who trod the mercantilist road by establishing a German trading colony in the early sixteenth century in what became Venezuela. Perhaps even more indicative of the externalized rather than centralized interests of the place was the creation during the sixteenth century of the informal "Fuggerzeitungen" which were newsletters for private commercial intelligence to aid the Fugger family and their clients in conducting trade that ranged from the Turkish marches to Antwerp and from the North Sea to Cadiz, and farther to South America. In the face of such reality, it is hard to believe that the present settlement pattern can be explained through retail gravitation in the 1930's.

Augsburg Needs an Externalized Model

It seems that even in the hearthland of central-place studies, with its specific conditions and special theory, it is reasonable to argue that facts exist to support a second model of settlement. That model is the one predicated upon externalization rather than gravitation. *Externalization* is the simple notion that cities grow in relation to their long-distance ties, those that are carried on by abstraction and agency, fully as much as by retail gravitation. Thus, we can broaden Christaller's special theory to allow the Fuggers and the Fuggerzeitungen to participate in the shaping of Augsburg along with West German farmers driving cars designed for Hitler. Not all cities are shaped by such external forces but those that are form the most significant class of places, and one whose function is largely obscured in the Christaller mechanism. Jean Gottmann alluded to such places in his impressionistic notion of the "economic hinge," but it remains for the present to try to discern and analyze the mercantile model of economic organization and settlement.

Before sketching a mercantile-capitalist model of trade and settlement, two additional concepts must be integrated into those of central-place theory. These are the related ideas of *demand accumulation* and

[10] The title of a section in the Regional Part of Christaller's study. Ibid., p. 180.
[11] Ludwig Wegele, *Augsburg* (Augsburg: Joh. Walch, 1966), p. 1.

agency of trade, which were presented in the first and fourth chapters. The satisfaction of demand often must be undertaken over interstitial distances (between customer and producer) much greater than was typical of customer conflux during the feudal era in south Germany. A trading system extending over larger geographical areas must rest upon a willingness of customers to accept less than instant satisfaction, and the creation of a class of agents to facilitate trade. Those agents are the origin of the mercantile adventurer, who soon became the entrepreneur of the capitalist system. The agent's absence in feudal times was determined more by the political system than any geographical order, as in many instances trade was physically possible when still officially restricted.

The contributions of the entrepreneur are several. First, he must exercise *judgment as to the location of a demand* that is not rigidly confined within the cells of an earlier feudalism and, thus, may be satisfied from several competing *mercantile cities*. The judgment required to predict the destination of men's trading desires is real, and subject to the reward of profit for correct evaluation. This is the reward for discerning *place utility*. Second, the merchant must appraise how large the demand accumulation has become, and plan to satisfy it. To fail in that judgment can result in unsalable surpluses or customers denied satisfaction, which in turn opens the market to successful competition by other traders, possibly even from other merchant cities. Successful entrepreneurs may be located either in the same place as the original merchant, who failed, or in different places because of the latitude of locational choice introduced through the agency of trade. With a greater latitude of location, the successful merchant must appraise both place and time utility.

The third contribution of the merchant comes along a quite different line, that of pioneering trade. So much has geographical thinking in these matters been shaped recently by central-place theory that the merchant is regarded as a primordial slug fattened or emaciated by the mechanistic operation of customer conflux, over which he has no control. On the contrary, the merchant has considerable ability to shape his own ends.

Adam Smith showed greater understanding when he noted, "The colony of a civilized nation which takes possession either of waste country, or of one so thinly inhabited, that the natives easily give place to the new settlers, advances more rapidly to wealth and greatness than any other human society." [12] Starting in the middle of the sixteenth century, European nations began to adopt the economic doctrine of mercantilism, which argued that a nation could gain wealth more readily by the conduct of trade than by the plowing of fields, or perhaps even the mining of precious metals. Acceptance of such an economic policy effectively freed the wholesale merchant from the rigid geographical restraints of the Middle Ages, with small trading units immured by

[12] *The Wealth of Nations,* bk. IV, chap. VII.

restrictions and vested practices, and gave him the ultramarine world as a field of exploitation. If it were not so, the world today would be very different because those merchants showed a level of enterprise which shaped the economic life of the two centuries before the industrial revolution and created the livelihood and political structure of much of today's world. The "New Lands" of our time are the plantations of the entrepreneurs of the commercial revolution of the seventeenth and eighteenth century.

By projecting demands over great distances and planting colonists to reap the products that were desired, merchants took hold of their own fate, and, doing so, came to be active agents in the creation and support of towns. The nucleated place totally ceased to be agriculture conducted from the town, as at Dinkelsbühl, but rather it became the world in microcosm. The death of feudalism had become the birth of mercantilism, and with that came a geographical transformation that our settlement theory has not yet acknowledged.

The Mercantile Model of Settlement as the Paradigm for Wholesale Trade

The job of the wholesaler is distinctly that of appraiser of market conditions. However much he may serve as warehouseman, deliveryman, and salesman, the contribution that gains the most for the merchant is correct prediction of the locale, time, and size of the future market. To maximize his return, the merchant seeks to organize that market over as large an area as possible.

The vigorous mercantile entrepreneur of the seventeenth and eighteenth century had to turn outward from Europe because the long history of parochial trade and the confining honeycomb of Christaller cells that had grown up with feudalism left little scope there for his activity. With overseas development, for the first time the merchant faced an unorganized land wherein the designs he established furnished the geography of wholesale-trade location. By contrast, in a central-place situation (such as that affecting much of Europe and the Orient), to introduce wholesale trade meant to conform to a settlement pattern that was premercantile. Thus observations about the location of wholesaling in a central-place system would be little more indicative of real choice than would be a geographical study of the residence of convicted criminals who had yet to complete their punishment. And in some cases, as at Augsburg, even in a feudal cadre there could be clearly mercantile places, dominated by external support, rather than retail center-seeking.

THE INTELLIGENCE COMPLEX. Save in such chosen-instrument situations as those of the monopoly companies, with their assigned territories, there was always the potential for competition, even in the conduct of overseas trading associations. Certainly the West Country towns of England—Plymouth, Bristol, and Exeter—all thought they should participate in the North American trade. But we should not be led into looking upon this as some strange variant of the central-place hierarchy. Quite the opposite, these towns had a structural interrelationship, but one of shared *knowledge* and *experience* rather than of a cellular market. To understand the tendency for a number of closely adjacent towns to enter the same wholesaling trade—as in the West Country of England in the seventeenth century, on the New England coast from Boston to Damariscotta in the earlier part of the last century, or in those settlements along the middle Missouri River from Kansas City to Sioux Falls in the latter part of the nineteenth century—we must introduce the notion of the *Intelligence Complex.* The discussion of Augsburg has already suggested its importance.

We know that the West Country began England's contact with the New World, and gained thereby the first intelligence of conditions of trade to be expected there. Seamen for the earlier voyages were drawn from several ports so it is not surprising that information they brought back should ultimately be spread among the same group of ports. When trade beyond codfish derived from the Newfoundland fishery was sought, these men were in a position of advantage because they knew somewhat more of the world beyond England than did sailors in East Coast ports; the West Country merchants shared this advantage. Certainly geographical location cannot be overlooked (these West Country ports were closer to America than Hull or Boston), but not too much can be made of that propinquity. In the India trade London dominated not because it was closer (Plymouth was the English landfall on the route to India), but because the monopoly had been vested in London and her merchants held most of the vital intelligence when the trade was freed in the nineteenth century. And we have already seen how the "Fuggerzeitungen" supplied the intelligence upon which Augsburg's early prosperity grew.

POINTS OF INITIATION. The effect of the intelligence complex was not lost with the concentration of wholesaling into the internal trade of the United States or Canada. *We find clear evidence that in wholesaling there is a greater tendency for competition to emerge within the initially important city than in some advanced spot closer to an outlying part of the market.* The Chicago situation seems typical of many types of mercantile activity. Because Marshall Field proved very successful in developing dry goods wholesaling from Chicago, his competitors showed a strong predilection to set up shop there as well. What this says is that Field had shown Chicago to be a good place from which to carry on dry goods wholesaling for a large interior area, and that he had there assembled important trading intelligence that might be tapped by others who hired away his employees. When mail-order merchandising came a generation later, the same flocking together of Montgomery Ward, Sears Roebuck, Butler Brothers, and Spiegel took place. In either instance there is a sharp distinction between the retail gravitation envisaged by Christaller and the organized market *ab initio* of wholesaling, or its cousin mail-order selling. The introduction of competition in wholesaling is as much organizational as it is geographical. As a result *points of initiation* assume great importance in the shaping of a theory of wholesale location.

THE IMPULSE TO TRADE. The point of initiation becomes the fixed pole from which the search for trade development takes place either in the classic mercantilist situation or in latter-day wholesaling. The outward reach from that pole is measured by the strength of the *impulse to trade*. That impulse establishes its force in two ways: from the absolute mass of the potential market and from the accumulation of demand that individuals practice. A large market can enjoy nearly continuous demand satisfaction, whereas a small market will of necessity have to accumulate demand to reach the critically effective strength to bring about trade. In the original mercantile development, *the interval of trade* was usually measured in months, sometimes in years, so that miniscule markets could still have trade, even if only very occasionally. And the dual office of most merchants—collecting staples in distant places for shipment to the metropolitan country and returning payment therefor, through the forwarding of manufactures to the collecting frontier—allowed them to carry on trade when the entrepreneurial profit from the conduct of either function might not be sufficient.

What Is the Mercantile Model?

The substantial distinction between the central-place and the mercantile model focuses on the source of growth and change. We have already noted the strongly endogenic quality of the Christaller design in which growth can come only through enhanced demands on the part of local consumers. Even if we grant that augmentation of that support may grow out of greater extraregional trade, the measures of externalities under central-place theory are indirect and blunt. As such, a rather cumbersome mechanism is required to account for the existence of high-order trading towns. That mechanism cannot be discerned even yet in many New Lands, and is almost certainly missing in the pioneering eras in those places. To explain the creation and growth of New Lands we must construct a model of settlement structure wherein the source of change is external to the developing overseas area or external to the parochial retail trade zone in Europe. The resulting *exogenic system* is the best approximation of a theory of location for wholesale trade that can be produced from such a preliminary investigation as here presented.

POINTS OF ATTACHMENT. The outward push of the European economy that came with the mercantile revolution led to the establishment of *points of attachment* between the two continents. In a sense these became what Jean Gottmann has termed "the economic hinge" between North America and Europe, but that name suggests a parity between the areas that adjoined along that line, which certainly was not at first true. When Bristol-born merchants gathered cod off St. John's or white pine timber on the Coast of Maine they began the spinning of a strand, thin and brittle at first, that has grown over the years. Yet the locus of attachment has not changed as greatly as most other settlement features in the intervening 350 years. St. John's, Quebec, Montréal, Boston, New York, and

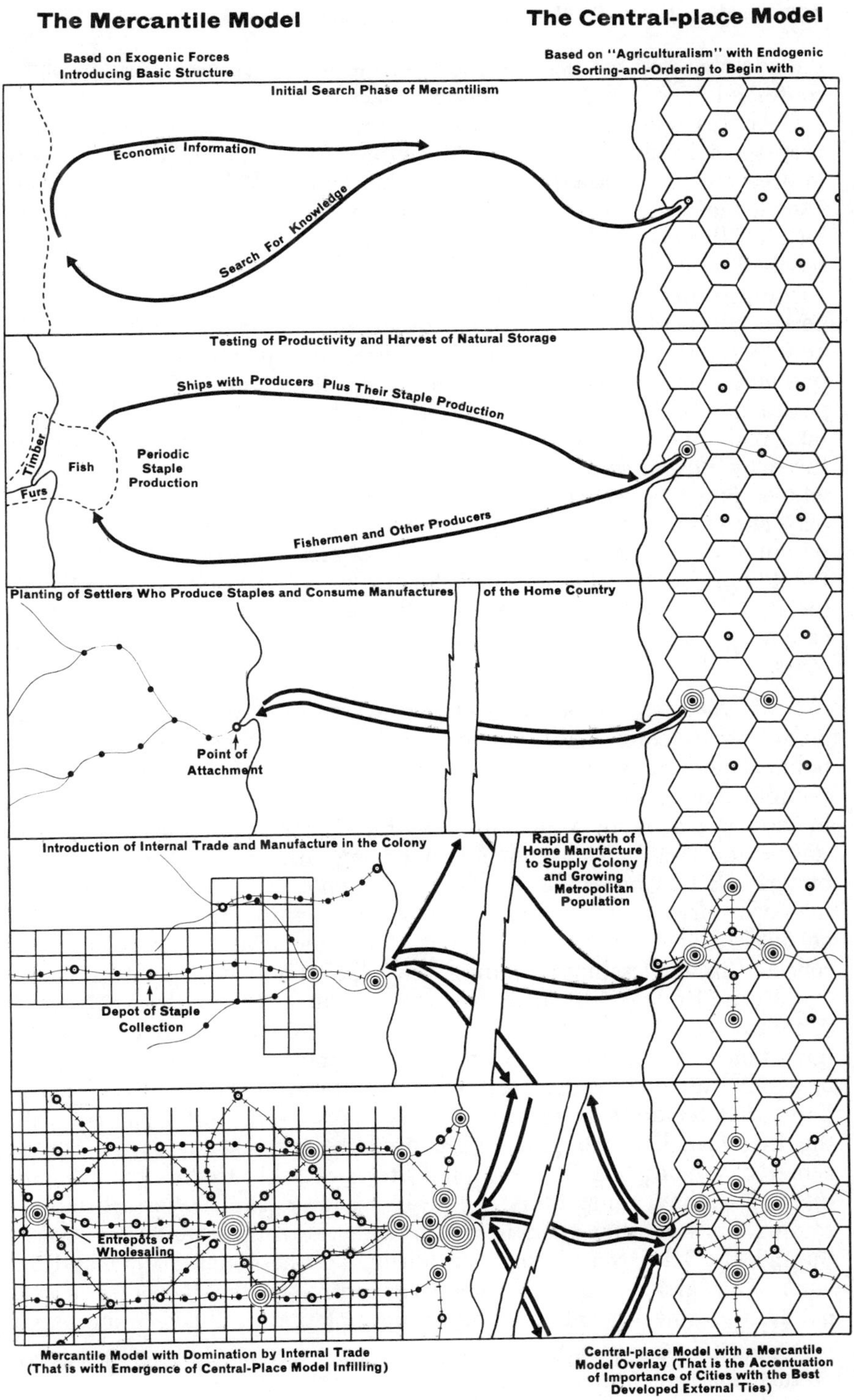

The Mercantile Model
Based on Exogenic Forces Introducing Basic Structure
The Central-place Model
Based on "Agriculturalism" with Endogenic Sorting-and-Ordering to Begin with
Initial Search Phase of Mercantilism
Economic Information
Search For Knowledge
Testing of Productivity and Harvest of Natural Storage
Ships with Producers Plus Their Staple Production
Timber
Fish
Furs
Periodic Staple Production
Fishermen and Other Producers
Planting of Settlers Who Produce Staples and Consume Manufactures of the Home Country
Point of Attachment
Introduction of Internal Trade and Manufacture in the Colony
Rapid Growth of Home Manufacture to Supply Colony and Growing Metropolitan Population
Depot of Staple Collection
Entrepôts of Wholesaling
Mercantile Model with Domination by Internal Trade (That is with Emergence of Central-Place Model Infilling)
Central-place Model with a Mercantile Model Overlay (That is the Accentuation of Importance of Cities with the Best Developed External Ties)

Figure 18

Philadelphia saw the tie established in the seventeenth century and have continued as remarkable wholesaling places ever since.

At this point it is well to recall that trade in its youth is seldom very specialized. The seventeenth-century merchant dealt widely, both in area and in product, and only as he gained massive experience and knowledge did he tend to drop the less rewarding lines and ports. The location theory of wholesaling needs takes account of this unspecialized seedbed to explain why a generally high level of wholesaling will tend to encourage the initiation, *in the same place,* of new lines of trade. The experienced merchant has the organization and the enterprise to undertake new lines, thus following the New England dictum that true entrepreneurs should "trye all ports" and risk all freights.[13] Thus, when innovation is under consideration, it is quite reasonable to hold that it will take place where entrepreneurial activity is most highly developed. In this way Boston was born on fish and naval stores but grew up as a wholesaling city on wool, leather, and electronic parts. Although cities may show some degree of specialization, it is usually as a component within a broad base; the single-function wholesaling center, such as High Point, N.C., is most exceptional. Such places are normally directly associated with localized manufacturing or adventitious entrepreneurial activity, as in the hardware wholesaling which dominates Fitchburg, Mass.

MULTIPLE POINTS OF ATTACHMENT SHAPE POLITICAL GEOGRAPHY. The points of attachment can be few or numerous largely in response to political and transportation conditions. The past practice of chartering a number of competing mercantile companies or colonies was reflected in the "Metropolitan Mercantilism" which grew up in the new federal state after 1789.[14] The competition among Atlantic ports to carve out hinterlands tied by canals, rail lines, and wholesaling links was the economic history of nineteenth-century America and continues to perpetuate itself. Even today in the United States, the several colonies are still reflected in wholesaling patterns, as may be true as well in Australia.

Even after the formal destruction of British mercantile ties in 1783, American economic history was shaped to a remarkable degree by the inheritance from the colonial times. The points of attachment continued to assume an independent status, and, in turn, began to project entrepreneurial tentacles across the Appalachians. New York's success with the Erie Canal, which probably assured the city's status as the largest center of wholesaling in the United States, quickly forced upon its *traditional* trading competitors the need to forge similar transportation–trading links. Boston tried to do so with a string of railroads which, through the *Michigan Central,* the *Burlington,* the *Union Pacific,* and the *Santa Fe,* carried the Hub's financial control to the Pacific. Unfortunately, the time lag between the opening of the canal-lake route as far

13 Samuel Eliot Morison, *A Maritime History of Massachusetts, 1783–1860* (Boston: Houghton Mifflin, 1961), p. 371.

14 George Rogers Taylor, *The Transportation Revolution: Industry, 1815–1860* (New York: Holt, Rinehart and Winston, 1951), p. 98.

west as Lake Michigan (1825) and the completion of the railroad to Chicago (1852) allowed New York City to entrench itself to a degree neither Boston nor Philadelphia has ever been able to overcome. Still, Pennsylvania, first by the Main-Line Canal and later by the Pennsylvania Railroad, attempted that task. Baltimore was early in the field with the Baltimore and Ohio Railroad, but too late in the battle to secure a victory. Farther south, Norfolk and Virginia made a start with a canal to Buchanan, and Georgia tried an end run with a railroad to Chattanooga. If there was any rationale to this feverish activity before the Civil War, it came from the already hallowed tradition of mercantile competitiveness. That New York won and few others finished is history, but an explanation of the *numbers* of public transportation projects strongly calls for a mercantilist model of settlement.[15]

INITIAL TRANSPORTATION SITES, THE ENTREPÔTS. The fundamental dynamic of the mercantile model of settlement is the operation of forces external to the local system, in sharp distinction to the endogenic conditions of the central-place model. Thus, most interior cities in the United States and Canada have not grown slowly out of the soil but, quite the contrary, were sited in terms of the trading function whose long external links were clear and full before the first lot was platted. Although most of these cities were "speculations," that is a totally insufficient explanation of their success in light of the "United States City" or "New York-of-the-Pacific" schemes that grew up to grass. Chicago was a speculation that succeeded just because it was well founded along an important line of trade at the lake terminus of the Illinois and Michigan Canal. Much as this is not the place to amplify the hypothesis, it might be noted that the "transportation city," which is so common in North America and so much less common in Europe, is, in most cases, not the outgrowth of the actual transportation function but rather stands as the fixed receptacle for the exogenic forces of wholesale trade. Beginning with steam navigation on the Mississippi system, as early as 1811, and expanding quickly with the opening of the Erie Canal in 1825, most of the receptacles were those critical to the operation of the inland navigation, which was the pioneering form of economic transport in North America. Buffalo, Cincinnati, St. Louis, and New Orleans are obvious examples but Cleveland, Toledo, Detroit, Milwaukee, St. Paul, Chicago, St. Joseph, and Kansas City are equally valid. Railroad building came merely as the technological improvement of conditions already determinate within an existing trading structure. As such, the laying of iron merely reinforced most previous conditions, although as Belcher has shown,[16] the shift did change some of the relative positions, as in the case of Chicago and St. Louis.

Depots of Staple Collection. The entrepôts, which grew up along

[15] See Carter Goodrich on the public transportation programs: *Government Promotion of American Canals and Railroads 1800–1890* (New York: Columbia University Press, 1960).

[16] Wyatt Winton Belcher, *The Economic Rivalry Between St. Louis and Chicago, 1850–1880* (New York: AMS Press, 1947).

the alignments of transportation projected inward from the coastal attachment points (Montréal, Boston, New York, Philadelphia, and Baltimore), were tied to those ports for two activities: The ports shipped onward to Europe the staples that became the economic products of the mercantile system in North America; and, initially, the ports were the place of importation for the manufactures and tropical staples which constituted the return flow. With such flows of staples these depots subsequently became the centers of manufacturing regions which, on independence, grew up as domestic substitutes for the English Midlands and the Scottish Lowlands.

We should not forget that wholesaling is as much a collecting as it is a dispersing function. Most of the mercantile cities of North America first waxed prosperous in the collecting trade—Montréal, St. Louis, Albany, and New York all dealt in furs; Boston and Charleston in naval stores; Cincinnati and Chicago in lard and pickled meat; and Toronto, Cleveland, and Buffalo, among others, in grain. These places were receptacles along the competitive transportation routes, for the conduct of wholesaling, and their specific location was determined by the proper place for the establishment of a *depot of staple collection* inland from the point of attachment.

The intelligence of conditions on the commercial frontier, which was gained in the collection process, became critical to the expansion of trading activity within the entrepôts to include the more traditional forms of wholesale distribution of manufactures. Thus in succeeding decades Montréal provisioned the fur trapper as Chicago the cowhand. Yet each began as the collecting agent for the staple gathered over half a continent. It seems, in fact, that Harold Innis's classic interpretation of the economic history of Canada as that of a trade in staples has wide validity when we seek to understand the growth of the entrepreneurial cities of all North America.[17]

Entrepôts of Wholesaling. Although most of the larger cities of North America began as depots into which the native staple could be collected for forwarding to mercantile partners, first in Europe and later on in the East Coast, these cities ultimately became the *entrepôts of wholesaling.* Merchants who began by collecting would have to change the details of their trade as the surrounding countryside developed. Still the spirit of entreprise would remain, even after forests were cleared and all the beaver trapped, and merchants would be most likely to go on to shape a new entrepreneurial activity with the knowledge and organization they possessed. *That knowledge related to the conditions of distribution and demand in the broad region tributary to their depot.* In similar fashion, the man seeking to set up shop for himself would be most likely to do so in a city where the mechanics of distribution would be at hand, and which was frequented by those customers he might seek to divert

[17] Harold A. Innis, *Essays in Canadian Economic History,* edited by Mary Q. Innis (Toronto: University of Toronto Press, 1956). Innis's "staple theory" of Canadian economic history is implicit in all these essays, but perhaps most succinctly stated on pp. 12–15.

to his new establishment. Thus, the shift from depot to entrepôt is a natural but highly important one.

STRUCTURAL CHANGE WITHIN THE CITY. The only element of the mercantile model that remains to be explained is the transformation of the pattern as a result of the elaboration of local settlement. *The basic assumption within the mercantile model is that of a local expression of distant support.* The demands of trade and the accumulation of wealth in Europe led to the mercantile settlement of North America. Only with the elaboration of commercial agriculture, in the nineteenth century, and the creation of a manufacturing belt was there the growth of any clear settlement hierarchy, such as that which had preceded by several centuries the creation of a long-distance trade in postfeudal Europe. In America the emergence of an elaborate localized retail structure signalled the shift away from a pure mercantile model of settlement. But we must note that the pattern was that of retailing implanted on an older mercantile structure rather than the reverse, as Christaller would have it. Even so, in the matter of wholesaling the shift was not too fundamental. The firmly established lines of trade, the entrepôts that had been designated, the fixed organization of intelligence collection, and experience and economies of scale, necessary to keep the cost of distribution down, all worked to preserve the pattern of earlier times in wholesaling. Only in the border zone between wholesaling and retailing, in the furnishing of goods to small artisans and service workers or to undercapitalized service establishments, was there much change as a result of population growth and the rapid urbanization of America. In most cities wholesaling spread into the outlying sections to gain the technological advantages to be secured by motor-truck transport. This spread was distinct from the creation of small stockroom wholesalers in those same outer areas, but the two joined together to influence the geography of wholesale trade location *within* cities to a degree that the economic development of the nineteenth century had not. Thus entrepreneurs, always ready to change for profit, have moved into the urban penumbra but they have not abandoned the vicinity of the "suns" which were designated early in the course of North America economic history.

Transformation of Model: Variable Factors

On historical principles we may discern the probabilistic frame of the mercantile model of settlement and the insight it affords as to the theory of wholesaling location. The points of attachment, the depots of collection and their succeeding entrepôts of wholesale trade in the New Lands can be envisaged and identified in actuality. Still two major questions remain to be answered: (1) In what geographical manner does the trade thus envisioned transform the earlier nonmercantile settlement of the metropolitan country, which initiated trade for national economic growth; and (2) In what ways does the mercantile model, even within the New Lands, become transformed on the maturing of trade and settlement in that area? Essentially these questions seek to establish the dynamic qualities of a system which is at root evolutionary.

The search for an answer must begin with these characteristics of the system that can vary with development, because any analytical system based on historical evidence is susceptible to an historical rigidity only slightly less confining than the straitjacket of assumptions of "natural order." The bind of history is the belief that the past is complete and unchanging, which is perhaps technically true, but there should be no doubt that the *impact* of the past is continually shifting with the varying composition of the present. And each adjustment that we inherit must be carried forward as well.

For example, the correspondence in time of the eclipse of the American merchant marine and the growth of domestic manufacturing caused several of the classic trading towns to shift from foreign venturing to factory manufacturing. Providence, Salem, and New Bedford each followed this course and the subsequent role each played shifted the city away from wholesale trade. In seeking to explain why, we may begin to discern the *variable factors* within the system.

The collapse of foreign trade quickly shut off the supply of information about potential markets. No longer was there a class of Yankee skippers who could quote for you the more recent prices for ginseng in Canton or the best sources of cocoa in Spanish America. And with that loss went the awareness of what goods would best buy those staples. Only the American market, protected by laws of cabotage, remained both open and known to the New England merchant, and that domestic commerce became attached to the cities that had forged interior ties by canal and railroad in the system of metropolitan mercantilism effected through wholesaling. The basic rules remained the same, but their impact could vary. The general truth persisted that commercial intelligence flow was critical to the practice of trade: we may call this the first of our variable factors.

This factor may be stated as a principle: The wholesale trade of any city (entrepôt) will extend outward as far as the limits of commercial intelligence available to that city's merchants. Extension of the intelligence limit will tend to expand the wholesale trade area, whereas contraction thereof will tend to reduce the trade area. The peripheral extent of the trade area is not necessarily uniform and the greater spread will occur along lines of best intelligence flow. Improvement in intelligence flows will tend to expand the trade areas of the larger wholesale center at the expense of the smaller. In a significant way this conflux of intelligence on the mercantile metropolis represents the geographical reciprocal of the concepts of "innovation and diffusion." With the ability to enlarge the intelligence field focussed on an entrepôt, which comes with improved communications, the traditional trading structure is able to maintain itself.

The second variable, which we can see from a careful appraisal of the history of trade, relates to the time required for demand satisfaction. With the collapse of American foreign shipping, or with the closing out of mining booms in frontier areas, the lag between the appreciation of demand and its satisfaction from distant cities would tend to increase. Obversely, the opening of railroads would tend to reduce the delay for supply from more distant entrepôts. These transformations would have intense influence on the lines of trade, encouraging clear shifts. Stated succinctly, it might be held: Increases in the speed of demand satisfaction are a constant objective of trade but one dependent upon the scale of demand and the technology of supply. Increases in the scale of localized demand lead either to forward dispersion of wholesaling facilities or to technological improvement in the distribution of goods from existing centers. Thus, technological improvements in transportation, as distinct from information communication, may also provide weapons to allow the traditional trading structure to maintain itself.

The third variable in the geography of wholesaling relates to the product itself. We have tended to discuss trade goods as if they are uniform throughout time, when we know that there are introductions and deletions in each epoch. The collapse of the sperm-wax candle trade on the introduction of kerosine, in the early nineteenth century, destroyed much of Providence's trade in the same way that the development of petroleum trading elevated Cleveland along with the Rockefellers. Intelligence of both supply and the market encourages concentration of trading, but the destruction of traditional sources of supply may well cause a sharp geographical shift in the wholesaling center. Once supply innovation occurs, the conditions obtaining at the beginning of any trade are reintroduced. Thus, a new set of supply-intelligence flow-demand conditions is established, and in such a context the shift of the "illuminating industry" from Providence to Cleveland may be understood. Clearly, improvement in neither communications nor transportation is adequate to maintain all traditional trading structures against all shifts in staple supply or consuming market.

The Entrepôt Alignment on the Geography of Wholesaling

Given the characteristics of entrepôts, in combination with the attractive force exercised by sources of staple supply, we may begin to explain one of the striking features of the specific geography of wholesaling in the United States. That *coup d'oeil* shows alignments of wholesaling towns of various sorts, but with a startling linearity that is totally unpredicted by the central-place model of settlement.

The first alignment, both in time and in importance, is that along the East Coast which comprises Boston, New York, Philadelphia, Baltimore, Norfolk, and by extension perhaps, New Orleans. These are the points of attachment between the massive closed system that is American domestic trade and its European hearth of development. Within that attachment this alignment is the very essence of the mercantile model of settlement, standing as a string of originally competitive footholds whose location was predicated entirely on external ties, rather than upon domestic trade and its geography. Only as the role of the staple collection changed and merchants sought *substitute trade*, which came to lie more in distribution than in collection, did these points of attachment become general wholesaling cities. Even so, they remain as the points of initiation of *American* external trade, as the ports through which we export the sort of goods Europeans originally shipped to us. The systemic qualities of trade are such that this turning around of the lines of flow should not surprise us, and any clear visualization of the mercantile model of settlement would immediately tell us that that reflux would be carried on from the points of attachment.

The second alignment encountered is that along the early routes of trade between the attachment ports and the staple-producing interior. As we have already noted, those alignments were shaped by the early technology of goods flow, by inland navigation, rather than by the more current overland rail and road routes. Thus, even in the interior where areal, as opposed to linear, arrangements of wholesaling towns might be assumed, the linear continues to be found. There is the *Lakes Alignment*, with Albany, Syracuse, Rochester, Buffalo, Cleveland, Toledo, Detroit, Chicago, and Milwaukee, and the *River Alignment*, with Pittsburgh, Cincinnati, Louisville, St. Louis, and Kansas City. Importantly, no such alignment is found along the Mississippi River, probably because it was not a route for the supplying of refluxive manufactures.

The rail-base may be seen in the *Piedmont Alignment* of the South. By the time there was a collection of cotton from the Carolina Piedmont, and a return flow of manufactures and food, the railroad played a major role, and the towns it encouraged stand as the main wholesaling places even today. Richmond, Danville, Greensboro, Charlotte, Spartanburg, Greenville, Atlanta, Gadsden, and Montgomery line up along the original rail route.

To some considerable measure, the supply-based alignment was at the origin of all these ranks of entrepôts, but only in the remaining sections of the country, dominated by the geography of staple production, is that alignment still found in its simple form. The edge of the manufacturing belt is so protuberant that its boundary is a zone, without any clear trading structure. In contrast, the edge of the truly staple-producing Great Plains is the home of an obvious *collecting alignment*. Minneapolis–St. Paul, Sioux Falls, Sioux City, Omaha, St. Joseph, Kansas City, Wichita, Dallas–Fort Worth, and San Antonio are its component trading towns. This is more an economic than a physical boundary, as would logically be the case for the present-day cutting edge of a mercantile system.

The linearity of the mercantile model of settlement is again reflected in the string of wholesaling towns lying at the western margin of those same Great Plains, both in Canada and the United States. *The Rocky Mountain Piedmont Alignment* from Albuquerque, at the southern end, northward through Pueblo, Denver, Cheyenne, Casper, Billings, Great Falls, Lethbridge, Calgary, and Edmonton forms at the physical break in terrain. Similarly, this economic flexure is an alignment of sites which possess, locally, the qualities desirable to pin down external ties of collection and distribution. Much as it might be assumed, following an areal paradigm such as central-place theory, that advance of the economic frontier would result in a cellular accretion of territories focusing on "trade centers," such a pattern fails to materialize in the North American West. In its place we find the continuing spread of systems established along mercantilist lines, and expressed in strongly linear patterns such as that along the Rocky Mountain Piedmont.

In the *Pacific Coast Alignment* the externally-originated string of towns is equally observable. The outside ties of the area were made *first by water*, and that fact can still be observed in the wholesaling geography. Much as the forwarding of Pacific Coast staples today is mainly to the remainder of the United States, and Canada to a lesser degree, with the compensating return trade from the same easterly direction, the alignment of wholesale towns is directly on the coast and highly marginal to the local markets being served. Only by recourse to a mercantile model may such a geographical pattern be explained. San Diego, Los Angeles, Monterey, Oakland, San Francisco, Portland, Seattle, Victoria, and Vancouver are wholesaling towns located by their initial entrepôt function and persisting, through the conservative qualities of the mechanics of wholesale trade. *History locates such wholesaling towns and tradition maintains them.* Only in situations of violent replacement, such as the substitution of the rail terminus Prince Rupert for the older coastal entrepôt Port Simpson, will the whole supply linkage be rebuilt on the coming of the transcontinental railroad. In this example from British Columbia, neither Port Simpson nor Prince Rupert, but rather inland Prince George with its better rail ties which led ultimately in three directions, became the wholesaling place.

A Systemic Argument

The broad evidence cited strongly supports an interpretation of facts rather different from those predicated on the areal assumptions of most previous work on the geography of commercial enterprise. No doubt the turned focus, concentrating on the wholesale component rather than the retail part of commercial enterprise, may help to explain why those facts call forth a different explanation "to determine geographic reality at present" from the one put forward by Christaller nearly forty years ago. In place of his *areal determinism*, with its endogenic change, we must advocate a regard for *system of trade*, with its exogenic change. This contrast shows up in the ability of the mercantile model to deal with open systems, to take account of external demands and productions as well as those in the parochial region, and to account for observable scalar similarity among cities based on quite distinctly contrasting *areal* supports. It is indeed in such a substitution that we can find the tools needed to explain why Boston and Philadelphia were in the front rank of English cities until the American Revolution, and why such a high proportion of the Australian population lives in a few overly large cities (when viewed in a central-place system).

It is unnecessary here to elaborate the details of those systems of trade, as this entire study is so aimed. Rather we reiterate that there are contrasting systems which have produced contrasting settlement patterns. Perhaps further study will support the notion that we deal with a number of special cases, with the *Heerschildordnung* of the feudal central-place model standing beside the *historical order* of the mercantile–capitalist model as two special case expressions. Even if such a pairing ultimately holds up, we still must substitute a systemic view of trade-settlement relationship for the simplistic determinism of bucolic trade conflux, which has backstayed geographical research in the lifetime of most practicing geographers.

The Mercantile Model of Settlement

At first it may seem strange to study wholesaling and to end up with a model of settlement, yet it should be apparent that the geographer's concern is not the mechanics of the economy but rather its physical expression. As such, the settlement of trade, rather than specifically its practices, is our concern. And, in the absence of full records of past practices and economic systems, we may use the physical evidence to expand our economic understanding, as distinct from our geographical awareness.

EXTENT OF SYSTEM. An economic system may be open or closed, in a geographical sense, depending upon the tendency to restrict or expand the field of operation. The central-place model argues that there are two expressions of economic activity—centralization and dispersion—yet if we examine these, we find each envisages a geographically-closed system, so there is no method for expansion within the system. Thus growth, which is so very much the rationale of mercantilism and, later, of industrialization, is not comprehended in central-place theory. To analyze these nonfeudal expansionist systems it has been necessary to introduce the notion of the open system geared to *externalization of the economy*. And, ultimately, we must add as well the comprehension of time.

THE MEASURE OF EXTERNALIZATION. Closed systems have been appraised in terms of the range of a good. What then is the measure of externalization? Specifically, the reach and articulation of the system. The *reach of trade* expresses the distance over which transactions are carried on and is, indirectly, a measure of the impulse to trade. For that reason the notion combines both the technological factors involved in transport and the propensity to accumulate demand. In truth, reach is an amalgam of both *act* and *ability*, the actual conduct of trade and the reserve capability to initiate trade given a sufficient impulse.

The second measure of externalization is found in the *degree of articulation* of the system. Most trade is not the simple coming together of buyers and sellers, even with the help of a lone agent, but rather the integration of a number of such linkages, which assemble into an articulated system. By looking at the physical expression of articulation, we may judge the functional nature of the system. Shifts in the location of articulation are good indirect measures of technological change and of changes in the propensity to accumulate demand.

DIGITATION AND CLENCHING OF ESTABLISHMENT LINKS. How great will be the reach from any wholesaling city? That *reach must extend from the place of wholesaling to the limits of the closed* (retailing) *system.* If the extent of the closed system increases, then the reach of the open wholesaling system may be contracted. The extent of the closed system refers to the mobility of men, mainly of shoppers. When men were fixed to the earth in traditional farming activities, they had to have trading centers close to their needs and shopkeepers whose level of selling was so low that they could hope neither for a sufficient business activity nor the time off to engage in "buying trips." To discover and develop new links to wholesalers was impossible. Instead, the wholesale traveller had to come to them, thus projecting the reach of the open system almost to the ultimate customer.

This relationship can still be seen in the supplying of goods to small builders and members of the various service trades. If we seek the most dispersed types of wholesaling, we find them in the supplying of building material (lumber, plumbing equipment, electrical goods), in the servicing of janitors, in business supplies, and in auto-parts wholesaling. In all these cases the wholesaler's customer is of sufficiently modest operating scale to require the wholesale firm to work hard to allow his stock-buying function to be kept to the minimum time and effort.

This need, which substitutes the wholesaler's mobility for his customer's immobility, leads to a fraying out into a number of disseminate wholesale trading locations, perhaps best viewed as *digitation* at the customer's end of the trade reach. The result is the creation of a number of distinct, localized, and closely adjacent trading establishments. It is important to note that such digitation is a location response to two distinct types of customers—to the highly parochial type associated with traditional land-tied occupations and to the local service trades. These occupations seem distinct but in truth are related by shared systemic elements. Along with much of the garment trade, some aspects of jewelry making and many other types of small-scale manufacture, servicing, farming, and the building trades, these occupations still hark back to the conditions which held in the early days of mercantile activity. Shortage of capital, the existence of a small and somewhat unpredictable "market," the absence of time to engage in purchasing expeditions, and many other qualities attach to these wholesaling customers. It should not surprise us that their patronage brings about the same geographical pattern of wholesale trade irrespective of the historical period.

Continuing the analogy, we must be aware that some small-scale artisans tend to cluster together, as in the garment trade. For them there will be a similar need to carry wholesaling far forward but there will not be the related digitation. Instead, *clenching* together of many wholesaling activities within the same restricted locale will take place.

A Geography of Staples, and One of Trade

Up to this point our discussion has, to sharpen focus, concentrated on the differences between the generalized central-place model and that related to the practice of mercantilism. The result is an overdrawing of distinctions, although there is little doubt that strong criticism of the central-place model can be constructive, and is overdue. The tendency of many urban geographers to treat Christaller's theoretical constructs as an all-encompassing explanation of the structure and role of cities must be dealt with; to launch a second model of settlement without dealing with the older and more widely accepted proposal would merely delay the intellectual resolution of the two models that must ultimately come. The question, therefore, becomes one of "universality" rather than of abstract truth. It seems completely unreasonable to hold, and unnecessary to argue, that central-place theory is without both abstract appeal and empirical validation *in places* and under certain *modifying assumptions.* It is the application of the system that is at issue. If central-place theory is viewed as the design for the explanation of a *special case,* then we must delimit the extent of its application and attempt to discern its relationship to the second model of trade and settlement here proposed.

When we examine the native soil of Christaller's thought, we can observe several strong correspondences to the abstract central-place theory. The theory has an innate "areal" basis, suggesting a tie between extensive economic production and Christaller's perceived patterns. In turn there is a tendency for such areal production to be related to the securing of staples, whose production creates large areas of economic homogeneity, facilitating the creation of hierarchial settlement structures. This dependence *upon the land* for economic support exaggerates the endogenic forces at the expense of those that are exogenic.

In summary form we may argue that, up to a point, central-place theory is an explanation of the trading-settlement structure of *an area,* particularly one of modest extent. For example, we might hold, as Brian Berry has, that Iowa's trading patterns grow within the Christaller model.[18] If we could view Iowa without its history or long-distance connections, the argument would be sound. Unfortunately we cannot treat even a staple-producing agricultural region in such an unreal way, except under extreme constraints of time and place. We must explain what led Iowa to produce its "staple" and where that staple goes once it leaves the farm. It is not sufficient merely to tell where the farmer goes in his Saturday marketing: we must also account for the destination of his hogs.

So long as parochialism is the economic rule, the central-place model suffices; but there is grave question that any place in North America

[18] Brian J. L. Berry, *Geography of Market Centers and Retail Distribution* (Englewood Cliffs, N.J.: Prentice-Hall, Inc., 1967).

CENTRAL-PLACE MODEL WITH ENDOGENIC CHANGE

BASIC FORM

Assumption That Placement Of Centers Comes From Conflux Of Retail-Service Customers And Subject To Political Or Speculative Ordering

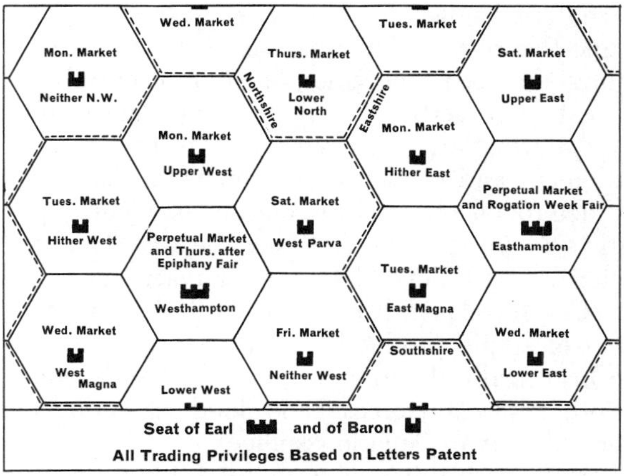

FEUDAL SORTING

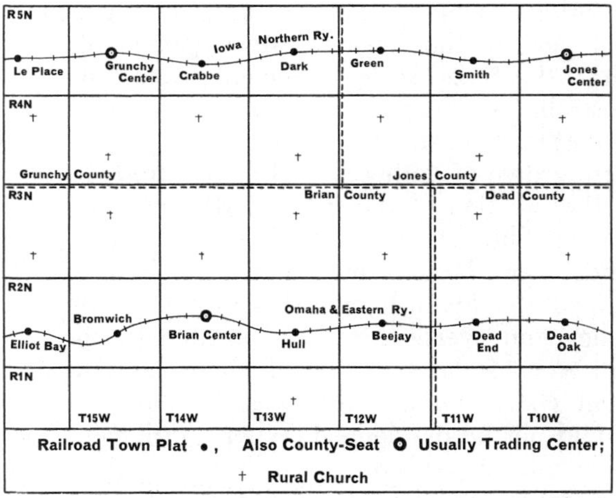

SPECULATIVE SORTING

EVOLVED FORM

Assumption That Rural Support Increases Due To Enhanced Agricultural Productivity And Growing Population

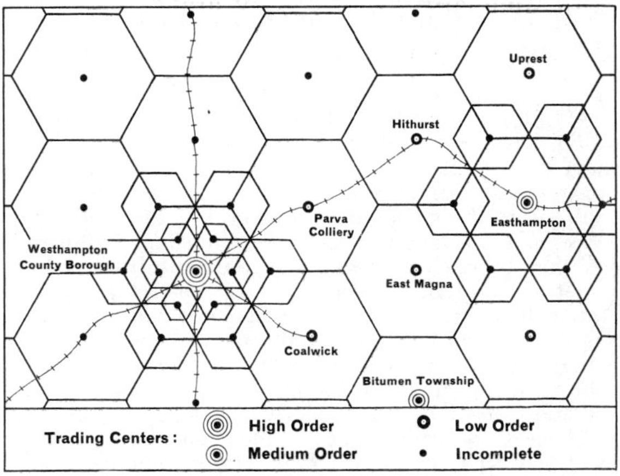

Figure 19

ever was sufficiently economically detached to have *begun* in a closed local system. In any event, once the externalization of economies took place, as signalled in particular by the adoption of the mercantile economic theories, explanation had to take on two levels of focus: that detailing the local settlement system (central place) and that detailing the extensive settlement system (mercantile model). This distinction is made particularly necessary by the observation that some patterns of settlement and economy are homogeneous and coherent whereas others are heterogeneous and discontinuous. Central-place theory can deal only with an homogeneity that is to be found in fairly localized occurrence. When we look beyond Iowa or Bavaria more is required.

Central-place theory may well be the geographical paradigm for areas of staple production, with the structuring of such areas reflecting the historically determined *parcelling-and-ordering process* of feudalism, or other similar forces.[19] So long as endogenic structure can be envisioned, this explanation suffices.

But if we think in terms of trade, a new geography is created. It is within that more extensive and heterogeneous areal structure of trade that most of the economic development of the New Lands took place. Thus, the broad-scale settlement pattern of Canada, the United States, Australia, and much of Latin America must be seen in the context of distant trade in staples rather than the production of staples alone. Staples were grown or gathered, but staple production was normally subsequent to the creation of the initial settlement pattern. Boston, Newport, New Haven, Philadelphia, and Charleston were planted before their hinterlands were hacked out of the wilderness.[20] The search for treasure shaped the Spanish city structure in Latin America whereas imperial policy and overseas trade delineated the urban geography of Australia and New Zealand. But in all these cases there was a mercantile settlement system before there was the staple pattern. This primacy in time should not be overlooked. Where settlement and economic development were introduced *from outside*, where there was exogenic growth, the mercantile model operates before the central-place, and its lineaments are the major ones of settlement. In Iowa, the Mississippi River towns and those along the Missouri grow out of the geography of American

[19] In the case of Iowa the parcelling-and-sorting force is not feudalism but rather the divisions introduced by the competitive railroad lines built between Chicago and Omaha (when that city was designated the official terminus of the Pacific railroad). Ultimately six essentially east-west lines were built across the state by as many companies. Using the geometrical base grid of the land survey, these companies founded "towns" as close together as seemed possible. Railroad builders held strongly to the belief that as much money was to be made from real estate speculation as from the actual transportation of goods. Thus, when the local staple economy of corn-hogs matured, there was already in existence a fine mesh of potential "trade centers" set upon a geometrical pattern by the parallelism of the Iowa railroads. From this grew the basic "Christaller system" that has been discerned. In that system there was no innate hierarchy such as is found under feudalism, and in Iowa the hierarchial aspects of central-place theory are poorly developed or are missing.

[20] See Carl Bridenbaugh's exposition of this thesis in his *Cities in the Wilderness: The First Century of Urban Life in America 1625–1742* (New York: Alfred A. Knopf, 1960).

trade, not out of the agriculture of the corn belt. Today they play a role in the livelihood of the corn belt, but to seek for their origin and location pattern in eventual parochial support is rather simplistic, *ex post facto* reasoning.

The staged quality of this relationship between the two models deserves emphasis. So long as we may deal with a closed economic system there is some support for applying the central-place model, regardless of time and stage, but once the system is opened up by long-distance trade, the new relationship must be taken into account. If the American case can be used as an indication of the nature of this tie, we might describe that case as broadly operating in the following manner:

1. Once the mercantile policy had been adopted, either by the English or the French, attempts were made to secure *preliminary information* about trading opportunities in the New World. In this stage we find the fact-seeking voyages of John Cabot (1497), Jacques Cartier (1534), and others.
2. Before settlement was attempted, the *economic potential* of the area was tested on the basis of information collected and the pattern of trade that might be anticipated. Such testing of potential may be periodic rather than continuous.
3. Once the potential was somewhat determinate, *initial settlement* took place in terms of a mercantile model. The dynamics were exogenic, the extent of the system was given by long-distance trading, and growth depended as much on the ability of the external world to consume as upon the parochial area to produce.
4. As the scale of the trading system enlarged, the mercantile towns grew, which in turn increased the demand for *hinterland provision* for those towns and for the *collection of staples* in greater quantity.
5. Only at this stage in the externally-based system did the *central-place model* begin to characterize settlement. That characterization was limited (1) to the later additions to the settlement pattern, and (2) to those areas characterized by areally-based staple production.
6. Subsequently there was *parallel growth* of settlement in accordance with both models. The mercantile model was the more broadly applicable because it dealt (1) with heterogeneity, and (2) with discontinuity. Only where (1) homogeneity, and (2) the continuity of areally-based staple production could be found was the central-place model likely to operate.
7. Each of these models was influenced by history. The scale and hierarchial form of the central-place model tended to reflect internal systems of parcelling-and-order. The scale and geographical extent of the mercantile system tended to reflect long-distance (external) ties established early in the history of the individual application of the model. Tradition worked to maintain retail ties in the central-place system and wholesale ties in the mercantile system.
8. In explaining the location of wholesaling it is necessary to concentrate attention on the mercantile model. In all likelihood there is a clearer explanation of retail trade found in the central-place model, as that design is specific to areally homogeneous and continuous systems.
9. The use of either model alone produces a simplistic abstraction rather than an explanation of the Geography of Enterprise.

10. Only in a "command economy," such as that in National Socialist Germany in the 1930s or the Soviet sphere today, can the single model serve, a fact that Christaller himself perceived clearly. He was led first to join the Nazi movement in the hope of reorganizing captured Poland along central-place lines and ultimately the East German communists to accomplish the adoption of the central-place model for the remaking of the German Democratic Republic's urban structure. If this tie between central-place thinking and what are imposed ordering-and-sorting forces was obvious to Christaller, it does not seem inappropriate to conclude that what it offered was not an explanation of economic enterprise but rather a Geography of Imposed Economic Order.[21]

Conclusion: A Study of the American System of Economy and Settlement

The nineteenth century was a time of unblushing critical comparison, unlike our era and its obsession with doctrinaire universality. Many Victorian essayists perceived contrasting solutions to common problems and were particularly ready to note "American systems" for accomplishing goals. Such was the early awareness of the geographical limitation of systematic solutions. The comparisons may have had an invidious quality when shaped by British writers but, overlooking that petty self-justification, they remain a good geographical analysis. Although he sought to avoid comparisons, Charles Dickens quickly came to view the America he visited in 1842 as "that vast counting-house that lies beyond the Atlantic . . . where everybody is a merchant . . . [and for whom there is a] national love of trade."[22] Well there might be, as trade had given birth to European settlement in North America and that trade had furnished the blueprint for the design of its economic geography. For the first time a vast continent had been occupied by men whose objective was commerce, in the pursuit of which they first devised an effective trading system to provide support and, later, created a form of settlement that can justly be called mercantile. The interweaving of the mercantile model of settlement and the growth of modern wholesale trade is fundamental to the extent that we cannot adequately explain the his-

[21] Christaller's long-continued desire to *put his theory to work* is attested to by Hans Carol. He tells us, "over a third [of Christaller's publications] are devoted to planning problems—to the way in which actual settlement patterns could be made to conform with the requirements of the theoretical model. As Christaller told me [Carol] in 1949, he lent his services to the Nazi regime in order to give advice about the creation of a hierarchical order of urban settlements for the newly won [sic] Polish territories. After the war he became a member of the Communist Party, for his hope was that an authoritarian regime would use its power to relocate war-devastated cities according to an optimal [sic] pattern as demanded by the central-place theory." "Walter Christaller: A Personal Memoir" (noting his death at Königstein, Germany, March 9, 1969), *The Canadian Geographer,* Vol. XIV, 1 (Spring, 1970): 68.

[22] Charles Dickens, *American Notes* (Greenwich, Conn.: Premier Americana, 1961). The quotations synthesize Dickens' view and appear, respectively, on pp. 43, 41, and 279.

torical economic geography of North America without great attention to the contribution of entrepreneurship and the wholesale trade.

Similarly, we cannot divorce the growth of wholesaling from an historical context. It has been customary in seeking to develop a general location theory of economic activities to assume a featureless plain over which these activities are distributed. Oddly, that absence of features is normally viewed as a terrain nullity when it would be more logical to view it as an historically vacant slate. In North America, for the first time, the trading economy could work out its destiny unstructured by confusing inheritances. Thus, here we are able to discern the true system of trade, and we find that the scouts become the wholesale merchants and the outposts of development are the settlements located and sustained by the international system of wholesale trade.